DAVE FUNK'S
TUBE AMP WORKBOOK

DAVE FUNK'S
TUBE AMP WORKBOOK
BASIC EDITION

Brig, LLC
8 Seaglade Circle
Keyport, NJ 07735
(646) 372-1853

FIRST EDITION
Eight Printing, June 2014

ISBN 10: 1500401064
ISBN 13: 978-1500401061

Front and Back Cover Designs by Mark Fair.

David Funk nor Thunderfunk Systems has connection or association with The Fender Musical Instrument Company.

DISCLAIMER: All information herein is for educational purposes only. There is no warranty, implied or otherwise, as to the correctness, accuracy, suitability, or merchantability of the information for any purpose whatsoever.

WARNING: Any and all changes made to electronic equipment must be done by qualified professional service personnel. Even an unplugged amplifier may contain Very High Voltages that are easily lethal even to an experienced electronic engineer. Do not remove the cover or chassis from your amplifier without professional assistance.

ACKNOWLEDGMENTS

I'd like to thank my wife Marjo, and our children for their patience with me during the last six months while writing this book, and the rest of the time as well.

To Ken Fischer, the builder of the finest sounding amps in the world, as well as a guy who can always answer the most difficult technical questions concerning guitar amps, and the tonal discordances of musicians.

To Greg Fritz for his expertise on speakers and Vox amps.

To Max Vague for his graphic design and computer expertise.

To Henry Kern at ACM and John Sizos at Minstrel Music for their help in building the first Thunderfunk amps.

To Michael Tackett for making Thunderfunk Amps and this book possible.

To Jody Carver for his early help in promoting Thunderfunk amps on the east coast.

To Brad Davis, Darin Favorite, Jack Pearson, Richard Aspinwall, and Mike Durham for playing and promoting Thunderfunk amps around Nashville.

To my buddies Bob "Norton" Thompson at Sound Check in Nashville, Joe & Paul Montineri at Montineri Custom Drums, and Fred Zadick at Fred's Guitars.

And to Al Greenwood and crew at Vintage Guitar Magazine.

PREFACE
2nd Edition

Well, here we are again. It's been 18 years and a lot has happened. I stopped building tube guitar amps and now only build solid-state bass amps, and have some studio stuff I'm working on. I stopped printing the First Edition of this book for no particular reason, and then it reached hundreds of dollars used at Amazon, I thought maybe the market could use a few more copies. This Second Edition also allows me the opportunity to re-write the book, adding information, correcting errors, and add a few new chapters.

Now back to the beginning:

When I started to write for Vintage Guitar Magazine three years ago, my intention was to write a book, a chapter a month. This forced me to write sequentially, starting at the beginning. The first year was frustrating, as I was just re-covering component basics that had been covered before. The second year we got into tubes, and this third year into the circuits themselves, covering material that I've never seen in print before. This book contains most of what was written earlier in the issues of Vintage Guitar Magazine, plus about 40% new material. There's some new information included here and pay particularly with what I learned about speakers. It is well worth re-reading in the new context, as explanations have been expanded, and additional graphics added to better illustrate the technical stuff. While the material is technical in nature, the reading level is 8th grade, making it easier to understand and remember. I'll try not to overly challenge you, or insult your intelligence, either.

This book will also serve you well as a reference to the history and authenticity of vintage Fender amps. The first chapter goes through the range of Fender products produced in the '40's through the '70's, identifying models and how they evolved. The schematics section can be used to date and authenticate amps you run across.

You can also use this book as a basis for studying vacuum tube technology, whether you're a musician or not. You'll be able to build a reference library that will help anyone better understand electronics, tube technology, and the history of vintage amplifiers.

And of course, It can be used as a service reference for Vintage Fender Amps. It was designed to open with a large schematic at the top and a large layout on the bottom for that reason. Schematic pages were created to fill in some gaps and I believe this is the most complete collection around. Please note that due to that the entire book is copywrited, INCLUDING the Schematics and Layouts.

INTRODUCTION

Enjoy the book, and I hope you learn something. I know I did by writing it. And don't forget the following warnings.

WARNING: To begin with, let me remind you that there are voltages inside an amplifier chassis that can reach 700 Volts Direct Current. Direct Current is not as forgiving as the Alternating Current found in the wall socket. Direct Current can clamp onto you and not let go, even after you're dead.

This book is provided for Professional and Educational purposes only, and is not intended to convince anyone, not previously qualified, to do work inside the chassis of an amplifier, EVEN IF IT'S UNPLUGGED.

Safety precautions will be explained and promoted, and CAUTION is given that if the information provided is used to service an amp, it is strictly at the user's risk. I cannot be sure that you understand all the possible exceptions that might be involved in a specific topic.

It's like the guy defusing the bomb, and his partner is reading him the directions. He starts with, "Cut the Blue Wire." After the wire is cut he reads the next instruction, which is, "But first..."

Just be aware, there's always more to be said on a subject. Reading this book is only the beginning.

If you make mods to your amp based on information I've supplied in this book, you're on your own. I do not sell kits, or plans to modify amps. I do not provide telephone "backup" for mods you do. If I wanted to say something, it would be in this book. As always, I'll be glad to take your phone calls concerning work I'm doing for you, or products that I sell.

This means that you'll need to have adequate troubleshooting experience to solve your problems. If you get in over your head, you're on your own. To help you with this, I intend to write a book on Troubleshooting Tube Amps.

I certainly cannot tell you that there is no danger involved; or that the reader is somehow qualified to service amplifiers just because he read this book.

If you are not personally, and absolutely POSITIVE of what you are doing

STOP

THINK IT OVER

and CONTINUE ONLY AFTER

you have DOUBLE CHECKED

what you intend to do.

Ready to start?

CHAPTER 1
A SHORT HISTORY OF FENDER AMPS

In 1877, Ernst Wermer, working for Siemens, invented the moving coil loudspeaker. The problem was, there were no amps to power it. It wasn't until 1906, when Lee De Forest invented the triode tube that amplification became possible, changing electrical engineers into electronic engineers. De Forest ended up working for RCA, after selling them his patent. RCA was using tubes for radio broadcasting, while Bell Telephone, and their manufacturing arm, Western Electric, were involved in audio amplification for lecturing. The first demonstration of amplified speech was in 1915, at the Panama Pacific Exposition, in San Francisco. This research lead to the development of patented circuits for constructing audio amplifiers, and by the 1920's amplified musical performances began.

Bell, Western Electric, helped Altec developed the 20-watt "Voice of the Theater" system, and in 1927 Western Electric sponsored the first movie with an amplified soundtrack, the "Jazz Singer."

At this time, the only way to get sound into an amplifying system was to use a microphone. The development of the electric steel guitar by Rickenbacker in 1931 lead to the development of other styles of amplified instruments, eventually changing the Spanish guitar from a jazz rhythm instrument into a Rock and Roll lead instrument.

In the 1920's, Leo Fender was a bookkeeper who got into ham radio as a hobby. This lead him to building and repairing radios for his friends, and renting out homemade P.A. systems. He eventually became a full time radio repairman, and opened a shop at Oak and First Street, in Santa Ana.

Early in the '40's, Doc Kaufman, a parts salesman and the inventor of the volume pedal, began to talk to Leo about electric guitars and amps. This lead to the K&F Manufacturing Company.

In 1945, the first K&F amps appeared with a gray wrinkle finish on a pine cabinet. Around a thousand of these 4-watt amps were produced, a few covered in tweed.

In 1946, Leo wanted to expand the product line with three new designs, but Doc hesitated, and then quit. This was the beginning of the Fender Electric Instrument Company, the fabled "Pre-CBS" company, with its first official amp being the Model 26, based on the original K&F amp design.

All these early amplifiers were based on Western Electric designs and patents, found in the Radiotron Designer's Handbook, and these amps had notices announcing "Licensed under U.S. Patents of American Telephone and Telegraph Company and Western Electric Company." By 1954 when the last of the patents lapsed, but Fender did develop patents of their own. The patents you've seen the most of are for the Tremolo (Vibrato) circuits in the amps. Most people think they're on the whole amp.

In 1947, Leo developed three new designs to go with the original Champ Amp. They were the Princeton, the Deluxe, and the Dual Professional, which in 1948 became the "Pro."

These amps started a schematic numbering system that I use in this book. The exception being the rule, the first schematic is actually the Bronco, which was put there for alphabetical reasons, and to break your concentration.

These new "TV-Front" Fender Tweeds were made from the summer of 1948 through the summer or fall of 1952. The name "TV" came about because of their resemblance to the front of a television set, with wide bands of wood surrounding the speaker grill. These amps usually have a piece of masking tape in the chassis with the name of the person who built it, or a date, or a "serial" number. I have seen a "5B3" Deluxe that was not a wide front, but this would imply that it was probably made in the first 6 months of '53, and that the "5C3" Deluxe didn't come out until the (then) big summer NAMM show.

So, we'll start with the #1 model amp, the "Champ."

THE CHAMP AMP - MODEL #1

The first of the "new" Fender amps was the 1947 Champion 600, with a 6" speaker, and a Champion 800 with an 8" speaker. The early ones were made out of uncovered hardwood, with red, blue, or gold velvet cloth. There were three vertical chrome strips attached to the front of the amp, to cover and protect the speaker.

In 1948 they were changed to a two-tone

cream and brown lacquered linen cloth. These had a small speaker opening with rounded corners, and plain brown, red (orange?), yellow, or blue cloth to cover the speaker. The control panel was located in the back, and slanted slightly upward. The name evolved to Champ 600, and, sometime in 1948, the covering was changed to Peroxaine, a trade name for luggage linen treated with nitro-cellulose lacquer, now commonly called "Tweed." This amp was designed specifically for steel guitar. Pretty funny when today steel players are using 1,000 watt amplifiers, and here's the smallest of the Fender amps being built for steel.

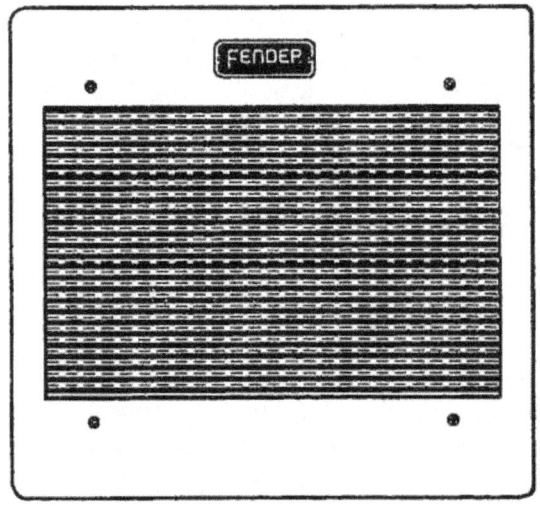

As mentioned earlier, the 1952 "5B1" Champs got a wider speaker grill opening.

The 1953 "5C1" Champ shows a single 6SJ7, octal based, pentode preamp tube, a single 6V6 power tube and a 5Y3 rectifier. Look at the assembly layout for the "5C1" Champ. Notice that the B+ voltage is taken off pin 2 of the rectifier tube. If you put an indirectly heated cathode rectifier tube into the amp, like a "GZ34/5AR4," you'll be pulling the DC Plate Voltage through the filament. This may not cause a problem in such a low powered amp, but it's something you should check. You could rewire the rectifier socket, but the loss of originality, and the fact it isn't easy to hide the rework, I'd say stick with the 5Y3.

Other books I've read said the Champ control panel was moved to the now familiar top rear position in 1948-49, but I've serviced a 1953 "5C1" Champ 600, and the control panel was still in the back position. The construction of this early Champ is interesting. A plain .060" steel chassis is formed by making one bend for the back panel. The chassis is PLATED with copper, not painted, as stated in another book. Sounds like something built by a radio man. The copper provided an easy surface to solder to, good for grounds, and excellent for shielding. Radiomen work in Faraday Cages — cages made out of copper plates, and copper screen. These cages isolate the engineers and the circuits they're working with, from outside radio interference.

The chassis had no sides or "front." Instead, it slid into slots in the wood sides of the speaker box, and two small screws held it in place through holes in the back faceplate. These wood slots support the chassis, in place of the missing chassis sides.

Sometime in late '53 or early '54, Champs received a dual-triode 12AX7 pre-amp tube stage. Although it now had two gain stages, they're triodes instead of a higher gain pentode.

Champs received sculptured grill cloth in 1955, and in 1956 all were changed to an 8" speaker.

The Champ's tweed was replaced by black Tolex in August of 1964, although the amps were still built in the "tweed style," with the control panel on the top rear. These would be rare, due to the short time they were produced. In January of 1965 the amp was changed over to the familiar top front "Blackface" position, and power was increased to 5 watts.

The "Vibro-Champ" was introduced in 1964. the BRONCO AMP is identical on the inside, to a Vibro-Champ. It was first made in September of '64 and discontinued in the early '70's. On the Silverface amps, the name "Bronco" is on the faceplate in red instead of the usual Fender turquoise.

In 1982 the Champ was redesigned and became the Champ II.

Fender also introduced the Super Champ version, with reverb. This amp used a Boost system of stealing signal off a 39K "Plate Resistor" on top of the reverb transformer, and mixing this signal back in while switching in another master volume control to compensate. What a mouth full. Add another 6V6 power tube in Class AB push-pull, and you have a resemblance to the original Champ. For an excellent collection of Jeff Veitch's color photos of vintage amps, get Aspen Pittman's, "The Tube Amp Book." As well as these beautiful color photos, Jeff is also a truly excellent black and white photographer.

Some of these Super Champs were made in Oak, and because of this there's a cool brown skirted Fender knob floating around the NOS market. I've used these knobs on some of my custom Nash amps, with cream vinyl covering, and a brass and brown marble face plate. Very cool looking.

THE PRINCETON - MODEL #2

The 1947 Princeton was similar to the Champ with an extra gain stage for a tone control, but still had 4 watts of power through one 6V6. It was designed for lap steel guitars. Early Princetons, had Princeton in script silk-screened onto the control panel. Princeton Street was five blocks from the Fender company.

The dual triode 6SL7 was changed to a 6SC7 in 1953 (5C2), and a 12AX7 in 1954 (5D2).

In 1957 the Princeton received a vibrato circuit.

The Princeton went blackface in 1963 with four black knobs. In 1964 the knobs were changed to white, and in '65 there were 5 knobs, with separate treble and bass controls. Power was increased to 12 watts by changing to two 6V6's.

August '64 saw the introduction of the "AA1164" Princeton Reverb Amp, although the schematic released is dated November, '64.

In 1982 Fender introduced the Blackface Princeton Reverb II. This Princeton has 29 watts of output, a presence control, a mid-boost, switchable lead, and a 12" speaker.

THE DELUXE AMP - MODEL #3

The Deluxe, with five tubes and an output of ten watts, was the top of the original 1947 single speaker lineup.

The original circuit had one 6SC7, a 6SN7 inverter — later changed to a 6SC7, two 6V6's, and a 5Y3 rectifier. It was the first Fender amp to use a push-pull output circuit, and to have two channels. One channel had one input, while the second channel had two inputs and a tone control.

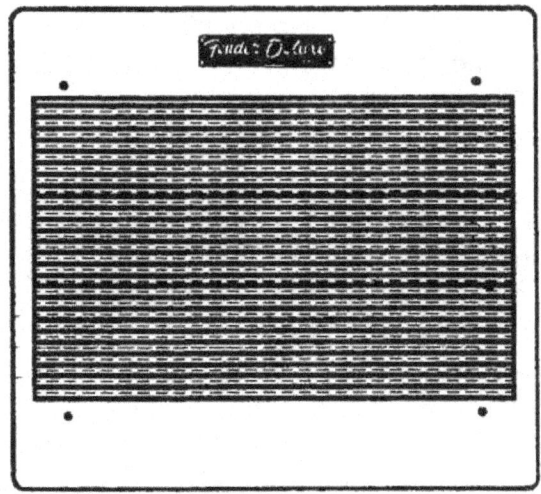

The 1952 "5B3" Deluxe is supposed to be a "TV" front, with wide sides around the speaker opening, but I've seen a "5B3" Deluxe built between the summers of '52 and '53, and it had a modern narrow front.

In the 1954 "5D3" model, the tube layout was changed to one 12AY7, one 12AX7 in a self-balancing paraphase inverter, two 6V6's with no feedback loop, and one 5Y3GT.

The 1955 "5E3" Deluxe was changed to two inputs per channel, one channel with tone and one without. The input 12AY7 stage was also changed from Contact to Self-Bias. This made the inputs DC coupled instead of AC coupled. Eliminating the input cap was a good thing. These caps can measure good, and yet block 75% or more of the input signal. If your Contact Bias amp is weak, measure an

AC test signal before and after the input cap, to make sure they're still working. Since they're on the AC signal side of the tube, instead of the DC power side, they don't cause DC power supply problems when they fail, making it harder to find the problem.

The 1955 "5E3" Deluxe received four inputs on two channels, still with a single tone control, and the inverter was changed to a cathodyne.

Another 12AX7 was added in 1957(?) for the tremolo circuit. It's been mistakenly reported that this added an extra gain stage to the amp. In fact, the extra triode half was used to change the inverter circuit from a Cathodyne to a Schmitt "long-tailed" inverter. This design stayed in production until the "6G3" was introduced in 1960.

The 1963 Deluxe became a blackface amp along with the Princeton, Pro, Twin, Concert, and Vibrolux. Separate tone controls were added to each channel

The Deluxe Reverb was introduced in August of 1963. It had reverb, vibrato, two channels, and 8 knobs. The amps puts out 21 watts into a Fender special design speaker. The vibrato channel was made slightly brighter, as there is no bright switch on a Deluxe.

In 1982, a new Blackface version of the Deluxe was introduced.

THE SUPER AMP - MODEL #4

Originally, the Deluxe was designed to be sold with the Broadcaster guitar, but in 1950 that honor went to a new version of the 1948-49 V-Front Dual Professional — the 1950 Super Amp.

The Super came with three 6SC7 preamp tubes, two 6V6's, changed from the Dual Professional's two 6L6's, and one 5U4G. The Dual Professional was the first Fender amp to use two speakers (10") to handle its 18 watts of output power, and the Super continued with this setup.

The 6SC7 tubes were replaced in 1954 with two 12AY7's and one 12AX7 tube. The 1955 "5E4" Super received separate treble and bass controls and power increased to about 20 watts.

In 1956, the power tubes were changed to 6L6's, and power increased to 32 watts. Tremolo was added in 1957, and power increased again to 40 watts.

In 1964 the Super Amp became the Super Reverb Amp, with four 10" speakers, 40 watts of power, reverb, tremolo, and tilt back legs. A new era had begun.

Solid state rectifiers were added in 1979.

THE PRO AMP - MODEL #5

The Pro is a one 15" speaker version of the Super which was simultaneously released in 1950.

It had three 6SC7 tubes which were replaced sometime during the 1953 "5C5" Pro model year by two 12AY7 tubes and one 12AX7 tube.

The 1955 "5E5-A" Pro received separate treble and bass controls and a presence control. The Pro followed the same development as the Super, with reverb and vibrato added to the "6G5" Pro Amp.

In 1960 the Pro changed to the front panel design, and was covered in Brown Burlap Tolex, with Light Gold grill cloth, and Dark Brown Knobs. It stayed this way until it went Blackface in January, 1963.

The "AA165" Pro Reverb Amp appeared in June of 1965, and is basically a 2-12" version of the Super Reverb Amp, or a half-power Twin Reverb.

THE BASSMAN AMP - MODEL #6

The first "5A6" Bassman appeared in 1951, to coincide with Fender's released of what today can only be called "THE Fender Electric Bass." This amp had two 6SJ7 pentode preamp tubes and a dual triode 6N7 (changed to a 6SL7 in the 1952 model "5B6" Bassman) as a paraphase inverter driving a pair of 6L6's for 18 watts of power, and a single 5U4 rectifier. The power supply design has what I can only call a defect. When you use a choke to filter ripple, you want to start off with a capacitor to ground. This has the effect of increasing the ratio of DC to AC. It makes some of the AC into DC, before it tries to snake through the choke. Since the choke resists AC but passes DC, the choke now becomes much more efficient. This might have been done to purposely lower the DC Plate Voltage. These amps did have screen resistors, in a way. They shared a common 10K resistor in what is today, the "choke" position.

The schematic shows a pair of speakers, but I've also heard that the amp had a single fifteen. I've never seen one of these amps but I have serviced the "5B6" Bassmans of 1952. They did have a single 15" Jensen speaker, and a speaker

port hole in the back of an otherwise sealed cabinet. Of course, to say sealed doesn't account for the fact that the amp's chassis was located at the bottom of the speaker box, with a little cut out so you could see the pilot light and get to the "on-off" switch. The "5B6" amps have the chassis on the bottom of the speaker cabinet, and the control panel at the top connected by an umbilical cord using an octal plug and socket for connection. This was a common approach, as the chassis, especially in such a high powered 18-watt amp, needed ventilation for it's tubes. The thing that scares me is the running of high DC voltages through the umbilicals of other designs. Fender only runs the inputs, and volume and tone control signals through this cord. Which isn't to say that high voltage didn't leave the chassis on Fender amps. Notice on the 5B4 Super Layout, at pin 3 of the 6L6's, it says, "To O/P Transformer Plate Mounted On Speaker." To identify amps with speaker mounted OP Transfromers, look for a lack of speaker jacks in the schematic and layouts.

In the 1952 "5B6" Bassman, the cap on the first stage tube was increased from 10uF to 250uF to increase the low end response. Marshall later built their amps with one of these 250uF bass channels and one bright ".68" guitar channel. By strapping these two channels together, you can balance treble and bass response by how much gain you use from each channel.

A negative feedback loop was added from the output transformer secondary to the cathode of the driver tube. This was to reduce low frequency distortion in the amp, improving the fidelity of the electric bass system. The power supply was also changed, eliminating the "leading" choke discussed earlier. This raised the plates to 400 volts. Two 16uF power supply caps were used to reduce the amount of low frequency power supply hum in the amp, knowing that the EQ for a bass guitar would be emphasizing any low frequency noise in the system. I doubt it was for "less bass sag" since "guitar amp" engineers did not think in those terms in the fifties. Everything was very technical, hi-fi, and by the book. While I'm absolutely sure that Leo listened intensely to what the musicians were telling him, I can't imagine Bill Carson saying, "Make it squeeze like a tube of toothpaste." More likely, common sense said, "Get the hum out of the low end."

In 1954, the "5D6" Bassman was changed to two 12AY7's, dual 5U4G rectifiers for less hum and more power, both important in a bass amp. The choke was brought back, but at the same time the design devoed (de-evolution) by eliminating the dropping resistor between the Plates and the Screens. You never want the screens to become more positive than the plates. That is one guarantee a screen resistor gives you. And, if it does happen, the screen resistor helps to limit the screen current.

I should mention here that the lack of Screen Resistors in these amps is not really the problem some think it is, especially in a Class-AB Push-Pull stage with Beam or Pentode power tubes. In fact, the Fender Output Transformer specifications for these amps, makes note that the Plates and Screens are tied together.

The inverter was changed to a 12AX7 cathodyne style driving a pair of 6L6's for 32 watts of power. The speaker configura-

tion changed from 1-15" to 4-10" Jensen speakers. Four small speakers fill more of the area available in a square speaker cabinet. This was seen again later, when Fender used the square Yamaha speaker in the Bantam bass amp. The Bassman still only had two inputs and but gained separate treble and bass controls, and a Presence control.

The tone controls are interesting because they were for the first time, driven by a low output impedance Cathode Follower. I've never heard anyone say before why this was done, but I'll tell you now. A Cathode Follower, the second tube in the circuit, has a voltage gain of about 95%. This is caused by local degenerative feedback. That's right, feedback. And what does feedback do? It reduces distortion, and in this case, hummmmmm. That's right, "It's a bass amp stupid!"

The 1955 "5E6" Bassmans were much the same, while the "5E6-A's" got a bias adjustment by changing a fixed resistor, and an adjustment of tone control values.

The 1956 "5F6" Bassman was changed to four inputs, a high and a low, each on separate channels, a bright and a normal, with a single common set of "modern" tone controls. A middle control was added for the first time, and a presence control located at the bottom of the tone stack was also the injection point for the amp's feedback loop. The presence control, and the feedback injection point, were both moved to the standard location at the grid of the inverter's bottom tube in the "5F6-A" revision.

The second preamp tube was raised from a 12AY7 to a 12AX7. In the "5F6-A" Bassman, the gain of the second triode was cut in half by removing the 25/25 cathode bypass cap. This is the amp that is the basis of legend.

The inverter circuit was changed from a cathodyne to the "modern" long-tailed pair. I believe that between Fender, Marshall, and Vox, guitarists have become so familiar with the sound of this inverter, that most prefer it on that basis. As a contrast, the Orange and Hiwatt amps of today still use the cathodyne inverter, and is a reason they have their great throaty sound.

The rectifier tube was changed to a single type 83. This was revised in the "5F6-A" Bassman to a single GZ34 rectifying tube. The 4 Henry, 56 ohm choke was moved to the conventional position of between the power tube Plates and Screens. This is known as a "Parallel Feed" circuit, and its purpose is to reduce hum. One hundred ohm Screen Resistors were added, which were later changed in the "5F6-A" to the now standard 470 ohms. Plate voltage on both amps was 432 volts. The speaker cabinet was also made slightly wider.

These amps have two .02 tone controls caps, although there are versions with 56K, .1,. 02, and 100K, .1, .047 tone circuits. The one I prefer, and use on my amp designs, is the "Marshall" style of .02, .02, with either a 100K or a 56K. Lately I'm favoring the 100K.

At this pivotal point in amp history let me stop and say that I've heard stories about why and how the Marshall amp looks just like a copy of the Bassman, but really isn't. Let me say that a very reasonable story I heard years ago was that Jim Marshall started building copies of, let me just say "American" amps, because it was just too expensive to import them. If

you've ever freighted an amp from LA to London, I'm sure you'll see the point. These Bassman amps also had no patents on them, and I see nothing wrong in whatever Jim Marshall did.

In 1956 the power tubes were changed from the Consumer/Commercial grade 6L6G to the new Industrial grade Tung-Sol 5881. Tubes with four digit numbers are usually ruggedized versions designed to withstand vibrations in an industrial environment. It's like, would you put a 12AX7 in your airplane's radio, or a 5751, or at least a 7025? I'm confident that the reason for the change was to better withstand the bass vibration. I don't think a guitar player at this time would want to play through something as lowly as a bass amp, with no special effects. So, I doubt that this change had anything to do with "singing sustain."

The most important thing about a production amp is its transformers. Fender transformers in the '50's, and early '60's, were made by Triad. Transformers and speakers are covered elsewhere in their own chapters.

In 1960, Fender produced both tweed and brown Tolex Bassmans. In 1961, the amp was changed to the new "Blonde" Tolex piggyback design. The speakers were changed from 4-10" to a single-12" in the ported cabinet. In 1962, this was changed to 2-12" in an infinite baffle (closed back) design. The first two years the amp came with oxblood grill cloth, changed to yellow for 1963, and to yellow with gold thread cloth in 1964. This cloth is not presently available in the after-market for restorations, so be careful. The knobs changed from white to dark brown in 1963, although I owned a '64 Bassman that had white knobs.

The Blackface Bassmans came out in August of 1964, although the knobs shown in the catalog remained white. This is the most popular Fender head for converting to lead because the circuit has an "extra" preamp tube, plus another unused triode stage, just waiting to be used.

The "AC568" Silverface Bassman had a new, larger, speaker box. The Bassman, Bandmaster, and Showman closed back cabinets had never really been big enough, and sound pretty stifled. This could have been fixed by adding one or two inches to the boxes depth. Instead, Fender was looking for the "big amp" look from the front. In mid-1969, the speakers were changed to 2-15's. There was certainly enough room for them in this big box. I have seen a 1967 Blackface Bassman with the big speaker box.

THE BANTAM BASS AMP
1969 - 1972

The Bantam Bass is a smaller version (one pre-amp tube less) of the Bassman, introduced in 1969. It was designed around a trapezoidal plastic cone "polyplane" speaker bought from Yamaha. The speaker was hard for Fender to obtain, and the Bantam was discontinued. The Bassman 10 uses a conventional 4-10" speaker arrangement, and was derived from the Bantam.

THE BASSMAN 10 AMP

This is a closed back 4-10" Bassman introduced in 1972. It has a normal guitar channel which sounds good, and a spe-

cial bass EQ'd channel that sounds horrible, even for bass. My advice is to rewire the bass channel.

THE BASSMAN 100 AMP

The 1972 Bassman 100, is a 100-watt bassman, or a Showman built for bass.

THE STUDIO BASS AMP

The 185-watt Studio Bass Amp, was a bass version of the Super Twin Reverb. The closest thing to an SVT, this amp was made from 1977 to 1981.

THE BANDMASTER - MODEL #7

The late 1952 "5B7" Bandmaster Amp had one 15" speaker, like its sister, the Bassman.

The 1954 "5D7" Bandmaster was the first to have the combination of three 10" Jensen speakers.

The 1961 "6G7" amps were the first "piggybacks" after the one off '58 piggy-back Bassman. They were covered in rough "Blonde" Tolex, showing they were considered professional amps. Originally coming with one 12" speaker, this was changed to 2-12" in 1962, just like the Bassman.

In 1968, the Silverface Bandmasters had the speaker cabinet stood on end vertically, and increased in size. There might be some 1967 blackface Bandmasters with the larger cabinet. I did see a '67 blackface Bassman in this style.

The Bandmaster Reverb Amp was introduced in 1969. It was basically a Super

Reverb chassis, built as a head. My complaint about it was, the head was bigger than it needed to hold the reverb tank. This was when Marshall amps were taking over, and Fender was looking for a "stack" look. This amp is also the same chassis as the Pro Reverb.

In the same way the non-reverb Super, Pro, Bandmaster, Concert, and Vibrasonic amps all share the same chassis. The biggest difference is the spacing and number of preamp tubes due to the addition of a reverb circuit and transformer.

THE TWIN AMP - MODEL #8

The 1952 "5B8" Twin was the new top of the line Fender amp. It was initially presented in the summer of '52 as a mate for the Stratocaster, but the Strat was delayed in production until 1954, with a bridge problem. It came with 2-12" speakers, which became part of its definition, two cathode biased, low voltage 6L6's producing 18 watts of power, and a single 5U4G rectifier tube. The preamp consisted of a pair of 6SC7's, a tone gain stage of a 6J5, and another 6SC7 for the paraphase inverter. The Twin became the first Fender amp to have separate bass and treble controls. This amp was the first to have the non-TV front styling, that was slowly adopted across the line starting in 1953.

The 1954 "5D8" Twin became the first Fender amp to have a presence control. The rectifier was changed to two 5Y3GT's, and the preamp tubes were changed to three 12AY7's, and a 12AX7 in a modified paraphase inverter circuit.

The 1955 "5E8" Twin was redesigned with a cathodyne phase splitter, and the bass tone control circuit was reworked. The dual rectifiers were changed to 5U4GA's, and bias was changed to fixed. These amps put out 30 watts.

In 1956 the "5F8" Twin got four 5881 power tubes with 100 ohms screen resistors, and a single large four pin type "83" rectifier. Tone controls became "modern," and the tone driver tube was changed from a 12AY7 to the higher gain 12AX7. This amp puts out an amazing, for the time, 60 watts of power.

The "5F8-A" Twin had the presence control and feedback point moved from the tone control stack to the inverter, the screen resistors raised to 470 ohms, and a single GZ34 rectifier installed.

The 1960 "6G8" "Brown Twin" (gold grill, brown knobs), and the 1961 "Blonde Twin" (oxblood grill, white knobs) had the cool early '60's vibrato circuit that caused a frequency shift by creating phase shift through a differential pair. This circuit is heavy in tube use, using 2-1/2 preamp tubes, and it's no wonder it was replaced by the opto-coupler circuit of the blackface amps, which uses only one tube — but the sound has never been the same. The tone controls were also changed to a more primitive style. This was the first Twin to use a solid state diode string for DC rectification. Plate voltage increased to 430 DC with power up to 65 watts.

The "6G8-A" Twin had the presence control changed to the 25K style. I much prefer the previous 5K style. The tone control were also changed back to a modern style. Just as with inverter circuits, musicians become familiar with the sound of a tone stack, and it's best not to change it. This is true with rectifiers as well. Have

10

you heard anyone complain about the "white" or "black" Twins not having a tube rectifiers?

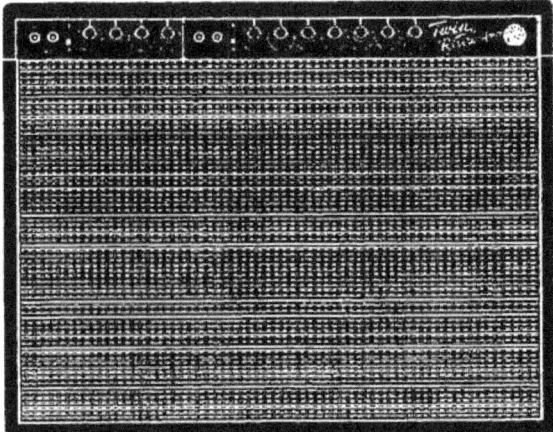

The "AA763" Twin Reverb Amp replaced the "Twin" in 1963 and a change in plate voltage raised the power output again to what I consider the standard for a Twin — 81-watts.

In 1963, JBL speakers became available, and these are the amps that I consider to be the top of Fender production. In the 1960's and '70's, our band would show up with a JBL Twin or two, a JBL Pro, and a JBL Dual-Showman for bass, and the people would always walk up and say in a whinny voice, "You're not going to play LOUD, are you?"

The 1968 "AC568" Silverface Twin, with the combination of self/fixed bias, 2,000 puff caps on the signal grids, and a bias balance control instead of a bias control, signaled the end of an era.

The Twin is basically the same amp as a Showman Reverb, Vibrasonic Reverb, Super Six Reverb, or Quad Reverb. The real differences are in speaker configuration, and therefore, output transformer impedance.

In 1970 the Twin got a master volume, and in 1973 the terrible push-pull distortion switch.

In 1976 Fender brought out the 185-watt "Super Twin Reverb Amp." It used six 6L6GC output tubes. In my opinion, sell it.

The 135-watt "Ultra-Linear" Twin was introduced in 1979. This design allows the screen grids to be operated at the new, higher, plate voltage of 500 volts. Normally, screens are held to 440 volts, even if the plates are at 500 volts. The "Ultra-Linear" design (meaning "low-distortion") allows the screens to be attached directly to the output transformer by using two additional "43%" taps on the primary winding. By doing this, power was increased from 81-watts for four tubes, to 112-watts. These amps are identified by their lack of a choke, and the huge laminate stack on the power supply transformer.

The 1990's saw a reissue of the "1965" Blackface Twin. While not as durable as an old Twin, these amps sound good. Because the amp is intended for clean playing, the problems that plague other reissue amps don't appear with the reissue Twins.

THE TREMOLUX AMP - MODEL #9

The 1955 "5E9" Tremolux was the first Fender amp with tremolo. It was a modified Deluxe designed as a mate for the Telecaster guitar. It was a 20 watt amp, using 2-6V6's, 2-12AY7's, a 12AX7 inverter, both sides of a 12AX7 for the tremolo, and a 5U4GB rectifier. It started out with a single 12" speaker. On other Fender amps, tremolo is called vibrato.

The 1960 "6G9" Tremolux had brown tolex, and a pair of 6BQ5 (EL84) power tubes. This amp must really scream. These were changed to 6L6's by the 1961 model year.

The 1961 "6G9-A" Tremolux became a white tolex piggy back amp, with a single 10" speaker in the Fender "surround" port configuration (see the section on the Showman). In 1962 an optional 2-10" version, in a closed back cabinet became available. The grill cloth was changed from dark brown to beige.

I bought my first Tremolux, a 1961 1-10" model, in 1972 for $75. I sold it 20 years later to producer Terry Thomas for $600. The only vintage amp I ever made money on, after inflation. Buying cheap is the secret. The Tremolux was discontinued in the middle of 1966 by the Vibrolux.

THE HARVARD - MODEL # 10

The 1956 "5F10" Harvard Amp came out in 1956. It was meant to be a mate for the Musicmaster guitar. When it first appeared, it had two 6V6 tubes, with a single 8" speaker. The "6G10" Harvard was later downsized to a single 6V6.

The Harvard remained virtually unchanged with the exception of changing to a 10" speaker. The Harvard was discontinued in 1960. In 1959 Fender did introduce a solid state version of the Harvard. Collectors hate it when I say things like, "The '5F10' Harvard screams because it uses a 6AT6 preamp tube. Go fight amongst yourselves.

THE VIBROLUX - MODEL #11

The 1955 "5E11" Vibrolux, was a cheaper, single channel with three inputs, version of the Tremolux. It was meant as a mate for the Duo-Sonic guitar, and came with a single 10" Jensen speaker. The Vibrolux used only one-half of a 12AX7 for tremolo. It was changed from tweed to brown Tolex, gold grill, and dark knobs amp for 1961, and the speaker increased from 10" to 12", although the catalog doesn't note this change until 1962. In 1963, both the Tremolux and the Vibrolux received the same chassis, with two 6L6's.

The "AA964" Vibrolux Reverb was brought out in August of 1964, and changed to 2-10" speakers. It had 35 watts of power.

The 1968 "AB568" Vibrolux had higher plate voltages, bringing its power up to 40 watts. The amp became silverface in 1969.

All the Vibroluxes make great blues amps.

THE CONCERT - MODEL #12

The "6G12" Concert came out in 1960 as a brown tolex amp. I was the same chassis as the Super/Pro/Bandmaster, except it had four 10" Jensen speakers, like the Bassman, but with a solid state rectifier.

Fender offered the same basic amp in four configurations: the single 15" Pro, the 2-10" Super, the 3-10" Bandmaster, and the 4-10" Concert.

The 1963 "AB763" Concert was a black-face amp. The Concert was discontinued in the summer of 1965. The Concert name was brought back in 1982.

THE VIBRASONIC - MODEL #13

The first of the new style Fender amps, was the late 1959 "5G13" Vibrasonic. It had 35 watts of power, and a single JBL 15" speaker was standard. It was basically a Pro, with Vibrato. Originally covered in rough blonde Tolex, oxblood grill, and white knobs, it was changed in January 1961 to the inverse with dark brown Tolex, light gold grill, and dark brown knobs. The Vibrasonic was replaced by the Vibroverb at the end of 1962, but came back in the middle of 1972 as the Vibrasonic Reverb — basically a 15" Twin Reverb.

THE SHOWMAN - MODEL #14

This was the amp everyone died for. For some reason, it had more sex appeal to me than a Twin Reverb, even though the Showman didn't HAVE reverb. I always knew I couldn't afford the $900 Showman (with JBL's, of course!), but when I heard a Fender Precision Bass being played through a Dual-Showman (tilted back of course), at the old McCormick Place (pre-fire) in Chicago, May, 1966, no problem. This guy filled the entire exhibit floor with that one amp, and I mean, he was shaking the place. I was there to demonstrate these new "Marshall" amps from England, but they didn't match the sound of that big Fender. We wondered why we needed 4-12" for guitar, and 8-12" for bass (the PA 100 amp), until we looked in back and saw the speakers. They were only good for 20-watts! No wonder! The English didn't know how to build good speakers! We found the tone to be hard, with little correction available from the tone controls. We played Stones, Beatles, Chuck Berry, and the Beach Boys. I left that day convinced that there was a Showman in my

future. I still have it. McCormick Place burned to the ground a week or two later.

The 1961 Showmans were among the first Fender piggyback amps, available with a single 12", or single 15" ported cabinet, covered in rough blonde Tolex, oxblood grill, and white knobs. The Tolex stayed Blonde, but the grill was changed to gold, and the knobs to brown for 1963. The speaker cabinet porting used a metal ring that held the speaker forward, but blocked the back of the speaker into a larger 15" or 18" hole, respectively. The larger hole was in a second, rear mounted, baffle board. The sound would travel around the back baffle to the side edges of the cabinet. From there, it would enter slot ports and travel between the front and rear baffle boards, exiting in a surround manner at the speaker. The port "surrounded" the speaker, and gave an apparent increase in speaker size. This same design was used on the single 10" piggyback Tremolux.

These heads are wider than the later Showmans, which had the same chassis size as the Bandmaster. The Showmans early output transformers were huge.

The "Double Showman" came out in December of 1962. The Double Showman became the Dual Showman halfway through 1963. The "AA763" is labeled "Showman," but the schematic shows two speakers, and the notes show a value change for the feedback shunt resistor depending on speaker configuration. The Showman didn't become blackface until August of 1964. The Single 12 Showman was discontinued at the end of 1965.

In 1968 the "AC568" Dual Showman got

the large (45-1/2" by 30" by 11-1/2")
vertical cabinet, and became a silverface.
The Single 15 Showman was discontin-
ued in the summer of 1969. The fact that
I've never seen a silverface Single 15
Showman is probably the reason.

The "AA768" Dual Showman Reverb
came out in late '68 or early '69, and the
non-reverb Dual Showman was dropped
by the end of '69. The Showman has
always been a piggyback version of the
Twin.

FENDER REVERB - MODEL #15

The 1961 "6G15" Reverb unit was
plugged in between the guitar and the
amp, and turned any amp into a reverb
amp. Reverb wasn't built into Fender
amps until 1963. The Fender Reverb was
reissued in 1994 and has sold very well. I
had the opportunity to examine my neigh-
bors Bill and Susan Carson's demo unit
shortly after its release. I found it to be a
very well built, exact copy of the original,
except for the use of a circuit board
instead of hand wiring, which I have no
problem with. One thing i really liked
about it was the 6V6 reverb driver tube
was run at the original 285 volts. This is
within the safe operating range for a
Reflector 6V6 (Sovtek) of no more than
300 volts. I'm confident that this unit will
give many years of reliable service.

THE VIBROVERB - MODEL #16

The first Fender amp with built-in reverb
was the "6G16" Vibroverb, released in
February of 1963. It was a reverb version
of the Vibrolux. Covered in Brown Tolex,
with 2-10" speakers, reverb, and 30-watts
of power, it was a screamer.

The release of the Vibrolux Reverb, in
August of 1964, caused the Vibroverb to
be changed to a single 15" speaker, with
blackface coloring. This is a great sounding amp, and my personal favorite,
although the 2-10" version is the more
sought after. The "6G16" Vibroverb used
a 7025 phase inverter tube changed to a
lower gain 12AT7 on the blackface models. The Vibroverb was discontinued
halfway through 1965, but reissued in the
'90's revival.

BLACKFACE - SILVERFACE

On January 5, 1965 the Fender Electric
Instrument Company was transferred to
CBS and renamed the Fender Musical
Instrument Company. When shopping for
an old amp, many times I'm told, "Yeah,
it's pre-CBS." At worst it's a Blackface
amp, and sometimes it's even Silverfaced!
Rarely, does it say "Fender Electric" in
small print at the bottom of the faceplate.
To straighten this out, it goes like this.

Fender Electric Blackface amps were offi-
cially produced from 1963 until January 5,
1965. Then the factory instantly converted all the faceplates to say Fender
Musical Instrument Company. Well,
maybe. I've seen Silverface amps with
AB763 tube charts in them, so who
knows. The first silverface amps came out
in 1968. I know. I went to Monti's Music to
buy a Blackface Bassman to match the
two Bandmasters already in the band, and
was surprised, and disappointed to find
the cabinet was bigger, and it had this
gaudy silver whatever you call it faceplate.
I never got used to it. Why did they do it?
The bigger cabinet wasn't too bad, except
when I had to move it, but that silver,

yuck! In 1972 I bought a 1965 Blackface Showman head that I still have today. I like the look of a blackface amp with black grill cloth.

Anyway, changing a Silverface amp back to a Blackface is a very simple operation. Change the bias control back to bias setting instead of the bias balance used on the Silverfaces. Remove the 2,000 pF caps on the power tube Grids (pin 5), and remove the 150 ohm resistors, and the 5/25 non-polarized cap from the power tube Cathodes (pin 8), and connect the Cathodes directly to the chassis (Ground). You're done.

CHAPTER 2
JENSEN SPEAKERS

The final stage of tone in an amp, for that matter, your entire audio chain, is the speaker. If the speaker's wrong, the tone just won't be there.

There are three basic speaker configuration for a Fender amp:

    1. Alnico.
    2. Ceramic.
    3. JBL.

JENSEN CONCERT SPEAKERS

The dark blue Jensen "Alnico 5" speakers in an open-back cabinet design, is half of the definition of vintage Fender tone. Throughout the early days, in the '40's, through the early '60's, most Fender amps came with Jensen Concert Series Alnico 5's. To quote the Jensen catalog:

"JENSEN Concert Series speakers have long been known and acclaimed by the trade and by users for their plus performance. From the earliest days, Concert speakers have been known as the finest speakers anywhere available for heavy-duty applications. Now, in greatly improved design, they are highly recommended for any purpose where exceptional power handling ability and high-quality performance are essential. Concert speakers are attractively finished in blue-gray lacquer and completely dustproofed."

"These PM speakers embody the highly efficient Alnico 5 magnets which insure long life and highest efficiency. Because Alnico 5 magnets are many times more powerful, ounce for ounce, than their predecessors, speakers so equipped offer obvious advantages: lighter weight, for savings in shipping costs; and smaller size, for savings in space in cabinet installations."

For comparison convenience, the Ceramic Speakers are include in the following charts. They start with the letter "C," while the Alnico speakers start with the letter "P."

JENSEN STANDARD SPEAKERS

According to the Jensen catalog, the "Standard Series speakers, although moderately priced, are exceptionally good in performance and are highly recommended for use in radio and television receivers, recorders, public address equipment, inter-communication systems and similar applications. Models listed have been completely redesigned in every detail. Magnetic structures have been designed to achieve maximum gap energy, cones selected for uniformity of response, and all speakers are completely dust proof. Standard Series speakers are finished in aluminum.

After comparing the Concert Speakers to the Standard speakers, three things become apparent:

1. There are no 6" Concert Speakers.
2. There are no 15" Standard Speakers.
3. The Ceramic Magnet Speakers have a higher power rating, at a lower price.

The column "ERG" is the "Gap Energy Level" in millions of ERG's. It is an indicator of the magnet strength, and this specification was not given with the Ceramic Speakers, probably to avoid direct comparison with the Alnicos. There are also no frequency range charts, nor frequency response curves in the Jensen catalog.

But how did the speaker evolve, and what's the difference between an Alnico, and a Ceramic Magnet?

THE EVOLUTION OF THE LOUDSPEAKER

The Loudspeaker, or linear AC motor, was invented by Ernst Wermer in 1877 while working for Siemens, the inventor of the rotary AC motor. (Recently, the "mho," or unit of conduction, an important factor in tube formulas, was renamed the Siemens.)

The problem was, there were no AC amplifiers to drive the linear motor. With the advent of the vacuum tube, amplification was possible.

In 1939, G.B. Jonas received a patent for a powerful new magnet material, Alnico. The name "Alnico" derived from its compositional material — ALuminum, NIckel, and CObalt, plus a little copper, and a lot of Iron. Cobalt and Nickel (which are chemically related to Iron) are magnetic, and can be used to make magnets more powerful than steel.

The Alnico used in speaker production is known as "Alnico V," or "Alnico 5," which relates to the balance of its chemical mixture, which is:

8 % Aluminum
14 % Nickel
24 % Cobalt
3 % Copper
51 % Ferrous

Alnico is an expensive material. Cobalt is a rare metal, obtained from countries like Zaire; Nickel is a metal so tough it's nickname is the "War Metal;" and airplanes are built out of Aluminum. When World War II began, private use of these metals, as well as most everything else, was restricted.

It's incorrect to think that only Ferrous substances, that is — substances that contain iron — can be attracted by a magnet. If that were true, then why is Stainless Steel, which is nearly three-fourths Iron, not attracted? The original magnetic substance, Loadstone, which is itself a variety of Iron Oxide, is an earthy rather than a metallic substance. Since World War II, when the countries supplying Cobalt began to raise their prices, the search for a replacement material for Alnico began. A whole new class of magnetic substances has been developed called "Ferrites." These are mixtures of Iron Oxides, and other metals, such as Cobalt or Manganese. By mixing with ceramic or plastic, and making a castable paste, magnets can be formed in almost any shape. By allowing the mixture to cool, and solidify under the influence of a strong Magnetic Field, the casting will become permanently magnetized, and a particularly strong magnet can be formed.

The Ceramic Magnet looks like a disc (or Cow Pie) clamped between the front and Back Plates of the Speaker.

the Gap, and restrict the Cone's movement to forward and back only.

SPEAKER FAILURES

Speakers fail for three reasons:

1. The Voltage applied to the Voice Coil is too high, forcing too much Current through the Voice Coil wire, causing it to overheat, and melt. Sometimes the heat only boils the varnish on the Coil, enlarging it, and causing it to short out between turns, or scrap inside the Voice Coil Gap.

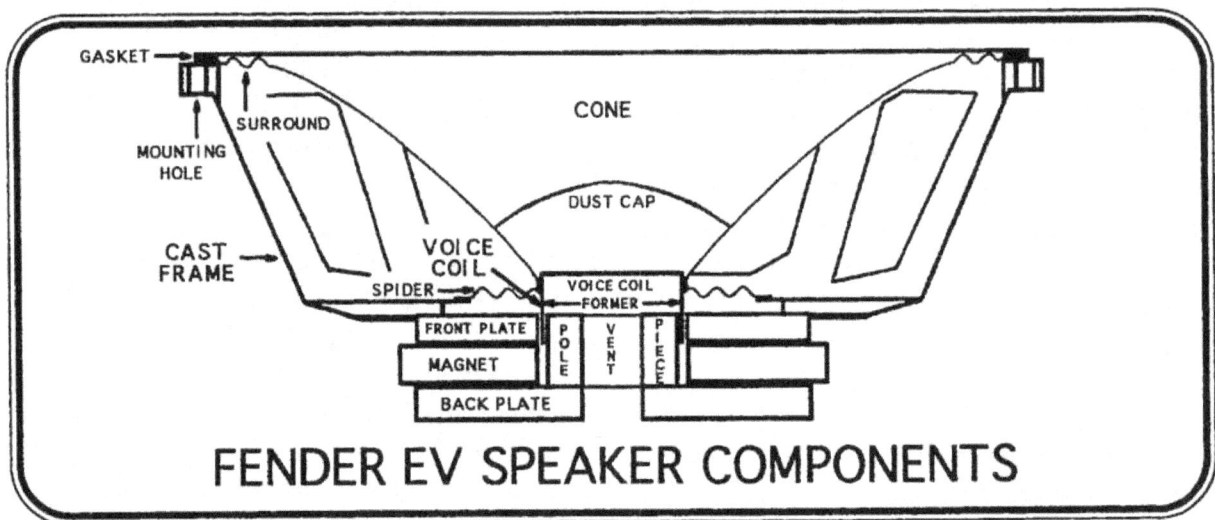

FENDER EV SPEAKER COMPONENTS

The Voice Coil is wound around a Voice Coil Former, which slides backwards, and forwards through the Voice Coil Gap. It does this by attracting, and repelling itself from the Magnetic Flux of the speaker's Magnet. The Voice Coil is held in center by the Spider. Attached to the front of the Voice Coil Former, is the Cone. The Cone does the actual work, and affects the sound of the speaker due to its Rigidity, Size, Excursion (Throw), and its Mass. At the front of the Speaker, the Cone is attached to the Frame by the flexible Surround. The combination of Surround and Spider keep the Voice Coil aligned in

2. The Cone is driven into excessive excursion, causing it to bottom out against the Backplate, or travel forward outside of the control of the speaker's magnetic field, allowing excessive Current Flow or mechanical misalignment.

3. Due to Mechanical Misalignment, the Voice Coil hits the Magnet or Pole Piece, damaging the Coil's wire.

The earliest speakers had 1/2" voice coils, and couldn't handle high wattage. When you laugh at a 1952, 18-watt Bassman, remember, these amps evolved from theater systems, and theater speakers are very efficient. When experiments to improve bass response lead to the acoustic suspension revolution, in the 1960's, speakers really began to eat power. While guitar speakers have always been efficient, the venues increased dramatically in size. So, from two directions our perspective changed. We expected 100-watts per channel home stereo systems, and 100-watt guitar amps. In the '60's no one ever thought they'd see a single speaker that could handle 400-watts. Today, there are companies trying to keep 1,000-watt speakers together.

Large voice coils dissipate heat better. The reason is obvious, they have more surface area. I owned a car with a Wankel engine, and it was like a BMW with two jet engines, but poor gas mileage. Efforts were made to increase the mileage figures for the Wankel engine, which lead to the conclusion that it could never match the mileage performance of a piston engine. Here's why. Its ratio of combustion chamber surface area to combustion chamber displacement is high. This means that the engine will always dissipate more energy as heat through the combustion chamber walls than a piston engine of equal displacement. The problem with speakers is, the more power you run through them, the hotter they get. The best way to dissipate that additional heat, is to install a larger radiator, as in a larger Voice Coil diameter. JBL's have Voice Coil diameters of 4".

But, small voice coils have better high frequency response. Maybe we should make the Voice Coil more heat resistant. We could use a plastic material, like Kapton, for the Voice Coil Former. Now it won't catch fire as easily. But Kapton doesn't sound right IN A SMALL VOICE COIL DIAMETER speaker. Where that leaves us today is, the MOJO 10" speaker is made with a small diameter voice coil, a paper voice coil former, but it will only take 20-watts. These are the closest thing today to a vintage Alnico P10-R.

SPEAKER ACCURACY, EFFICIENCY, & POWER HANDLING

Most people think that the single most important specification for a speaker is flat frequency response. The hold over from this belief is when you see a guitar speaker manufacturer tout "frequency peaks at all the right places." Actually, the edge of the sound wave transmits the most information about what we're listening to. Remove the percussive envelope from the sound of a piano, and you'll probably hear a flute. The idea is to get the speaker to move forward instantaneously, stop dead in it tracks, and then accelerate in reverse at full warp speed. This is called speaker "damping." It has to do with how well the amplifier has control over the movements of the voice coil. You might think of it as a lack of "ringing." The specification of speaker damping is the ratio of the amplifier's output impedance to the speaker's impedance. The lower the Output Impedance, the better.

The next way to improve speaker response is to have a larger magnet. Almost everyone says, "Wow, look at that magnet. This must be a really high power speaker!" Not necessarily true. Just as with transformers, the size of the transformer has more to do with frequency response than power handling. The same with speaker magnets. The larger the magnet, the more efficient the speaker. You might think of it as having a "more" solid surface to push off of. A weak magnet can't develop the push. The speaker will be mushier, and waste power in the translation. As you pump more power through the speaker, the voice coil motion becomes less controllable. Eventually, you overpower the magnet. This is the only reason the size and strength of the magnet makes a difference to power handling. The real definition of power handling is voice coil dissipation, as we've already seen.

The Flux Density of a speaker opposes the Voice Coil, so it makes sense that it's needed in the Voice Coil Gap. Alnico magnets are easier to focus the Flux Fields into the Gap because they have a greater Flux Density than Ceramics. Therefore, Ceramic Magnets have to be larger. Alnico speakers have their Magnets inside the Voice Coil. If you remove a cover from a Bassman's Alnico speaker, you'll see how the voice coil slides OVER the magnet. The Horseshoe Ring is there to focus the Flux into the Gap . A ceramic magnet is too big to put inside the voice coil, and you've probably seen the ceramic "disc" attached to the back of a ceramic speaker.

THINGS TO REMEMBER

Thanks to Ken Fischer, everyone now knows that Jensen are "Reverse" Throw. If you attach a 9 volt battery to the speaker terminals (I use my weak, worn out ones), Positive to Positive, the Cone in Jensens and JBL's, will move Backwards. Celestion speakers will move Forward. Remember this when mixing speakers

and solving phasing problems. To determine the Impedance of a Speaker, multiply the DC Resistance of the Speaker by 1.4 times.

JBL

James B. Lansing's first started manufacturing speakers under the name, Lansing Manufacturing Company. He later merged with Altec to become Altec-Lansing, and became a vice-president of that company. This didn't work out, and he left to form the James B. Lansing company, commonly called "JBL."

JBL started with field coil units, progressing to permanent magnets, and becoming the standard studio reference speaker in the days of Mitch Miller, and later Clive Davis, at Columbia Records. The reason I mention this is to remind you of the high standards of quality and consistency that Columbia gave to the record industry in the '50's, '60's and 70's. If you recorded for Columbia, it was at a company owned Columbia Studio. The highest standards of excellence were maintained. To name a few, Barbra Streisand, Simon & Garfunkel, Chicago, Blood Sweat & Tears, Bob Dylan, and Janis Joplin. This is what the JBL speaker was. A very smooth, "neutral" to the point of being flat, but still sweet sounding, and high power speaker. The large 4" Voice Coil, made out of Copper or Aluminum, gave it great efficiency, stability, and power handling capability. . Even though it was neutral sounding, it still had the Alnico "kick" to it. Since the large Voice Coil tends to limit high frequency response, an Aluminum Dust Cover was attached that almost acted as a Whizzer (a small cone attached to the middle of a speaker cone to enhance its high frequency response), significantly brightening the sound. With a cast frame, these speakers had tremendous power handling capability for the time. Whether you like them or not, JBL stills makes some of the finest speakers in the world.

As early as 1945, JBL had a suitable guitar speaker in the JBL-D101, but it wasn't until late in 1959, that Fender introduced the Vibrosonic Amp, with a JBL D-130 as standard equipment. In 1963 Fender discontinued the Vibrosonic, but then offered the JBL D120, D130, and D140 speakers as an option, in the Showman, and Twin amps.

The JBL "D" speakers were painted Gray, but the ones delivered to Fender were painted a fairly bright Orange. This is an easy way to tell them apart.

The JBL "K" series was introduced in 1973, as I recall. There were two rumors I heard at the time. One was that the purpose of the new design was to defeat the Nixon price freeze that was on at the time due to the start of rampant inflation, at the time blamed on the oil shortage, by redesigning, and therefore creating a new product that could be priced higher. The other rumor was that the power rating was increased by enlarging the Voice Coil Gap. This, of course, lowered the speaker's efficiency, so it would take a larger amp to drive it. I don't know if either of these rumors is true. If I knew what I was talking about, I'd be writing an entire book on JBL. In any case, the Orange backed "D" are the ones that players and collectors are looking for.

The next JBL speaker was the Model "E." These speakers use a Ceramic Magnet, with what JBL calls, a Symmetrical Field

Geometry. These "E" model speakers are said to embody the qualities of the older "D" and "K" Series Alnico speakers.

FIELD COIL SPEAKERS

To my knowledge, Fender never used a Field Coil Speaker, but since I'm talking about Jensens, and there's such a lack of information out there, here's the deal on the Jensens some guys swear by.

Before speakers were built with things called "Permanent Magnets," there were Field Coil speakers, that were basically "Electro-Magnets." They received their magnetizing power from a DC Current flowing through them in the place that a Choke would normally be found in the power supply. This presents the dangerous idea of running high DC Voltages out to the speaker and back. I don't like it, but there it is. Perhaps the most important specification here is the Resistance of the coil in ohms, to determine a replacement value if you change the amp over to a Permanent Magnet speaker. The Jensen F12-Q, and F15-Q are 1,000Ω, and the F12-N, and the F15-N were available as either 4,000Ω or 5,300Ω.

DATING SPEAKERS

For those guys that prefer dating speakers, this section is dedicated. On the back of the mounting rim, you'll usually find a six to eight digit number stamped in white ink. The first three (or four) digits are the EIA manufacturer's code. The EIA list below shows companies that may never have manufactured speakers, but it was convenient to list them all here, in one place. They are:

24 Becker
34 Cornell-Dubilier
67 Eminence
101 Admiral
106 Allen-Bradley
117 Credence Speakers
119 Automatic Mfg.
124 Alpha Wire
125 Bendix
130 Matsushita/Panasonic
132 Talk-A-Phone
134 Mepco/CentraLab/NA Philips
137 CTS
140 Clarostat
145 Cinaudagraph/Consolidated
145 Illinois Capacitor
150 Crescent
169 Hitachi
185 Motorola
188 GE
213 Dearborn Wire
220 Jensen/Viking
230 Littlefuse
232 Magnavox
235 North American Capacitor
245 National
251 Ohmite
252 DuKane/Operadio
260 Philco
270 Quam-Nichols
274 RCA
277 Emerson/Radio Speaker
280 Raytheon
285 Rola
286 Ross
296 Solar
304 Stackpole
308 Stromberg-Carlson
312 Sylvania

328 Utah/Oxford
336 Western Electric
343 Zenith
371 Best
381 Bourns
391 Altec-Lansing
394 Foster Transformer
416 Heath
433 Celveland
449 Wilder
465 Oxford/McGregor
466 Delco
549 Midwest
575 Heppner
579 Belden/Cooper
589 Bogen
649 Electro-Voice
706 Pioneer
719 Carbonneau
736 Sprague/Allegro MicroSystm
742 Esquire
748 Russell
756 Universal
767 Quincy
787 Sonatone
789 McGregor
794 Harmon Kardon
795 Atlas
828 Midland
840 Ampex
847 University
918 Oaktron
1056 Fisher/Gefco
1059 Channel
1098 Pyle
1113 Acoustic Fiber Sound
1149 Curtis Mathes
1191 Micro Magnet

The next one or two digits are the year code, and the last two digits are the week code. If you see 220348, it stands for: Jensen, the 48th week of the year 19?3. You have to decide if it's 1953, or 1963. By 1973 the code would read 2207348. If it were a Pyle speaker it might say 10988348, for the 48th week of 1983.

JENSEN CONCERT SPEAKERS

SIZE	MODEL	ERG	MAGNET WEIGHT	DEPTH	VOICE COIL DIA.	OHMS	WATTS	LIST PRICE
8"	P 8-R	2.2		4"	1"	6-8	9	$15.25
8"	C 8-R		10oz	3-1/16"	1"	8	12	$12.50
10"	P10-R	2.2		5-1/4"	1"	6-8	10	$18.50
12"	P12-R	2.2		6-1/16"	1"	6-8	12	$19.50
12"	C12-R		10oz	6-1/16"	1"	8	14	$16.00
8"	P 8-Q	3.2		4"	1-1/4"	6-8	10	$24.20
10"	P10-Q	3.2		5-1/4"	1-1/4"	8	12	$26.30
12"	P12-Q	3.2		6-1/16"	1-1/4"	8	14	$27.75
15"	P15-Q	3.2		8"	1-1/4"	8	16	$35.00
12"	P12-P	4.6		6-1/16"	1-1/2"	8	16	$40.00
15"	P15-P	4.6		8"	1-1/2"	8	18	$47.25
12"	P12-N	6.6		7"	1-1/2"	8	18	$49.00
12"	C12-N		27oz	6-1/16"	1-1/2"	8	20	$52.50
15"	P15-N	6.6		8"	1-1/2"	8	20	$55.00
15"	C15-N		27oz	7-1/4"	1-1/2"	8	25	$61.50

JENSEN STANDARD SPEAKERS

SIZE	MODEL	ERG	MAGNET WEIGHT	DEPTH	VOICE COIL DIA.	OHMS	WATTS	LIST PRICE
6"	P6-X	.25		2-3/4"	9/16"	3-4	3	$ 5.00
6"	P6-W	.36		2-7/8"	9/16"	3-4	3.5	$ 5.65
6"	P6-V	.51		2-15/16"	9/16"	3-4	4	$ 6.10
8"	P8-V	.51		3-3/8"	3/4"	3-4	5	$ 7.30
8"	P8-U	.74	1.7oz	3-1/2"	3/4"	3-4	6	$ 8.35
6"	P6-T	1.1		3-3/16"	3/4"	3-4	6	$ 7.75
8"	P8-T	1.1	2.5oz	3-5/8"	3/4"	3-5	7	$ 9.50
10"	P10-T	1.1	2.5oz	5-1/4"	1"	6-8	8	$10.65
12"	P12-T	1.1		6-1/16"	1"	6-8	9	$11.85
8"	P 8-S	1.5		3-13/16"	1"	6-8	8	$12.25
8"	C 8-S		6oz	3"	1"	8	11	$10.35
10"	P10-S	1.5		5-1/4"	1"	6-8	9	$15.25
10"	C10-S		6oz	5-1/4"	1"	8	12	$13.50
12"	P12-S	1.5		6-1/16"	1"	6-8	10	$16.50
12"	C12-S		6oz	6-1/16"	1"	8	13	$14.75

CHAPTER 3
BASIC ELECTRONICS

An atom is composed of positively charged particles called protons, and a much smaller negatively charged particles called electrons. The protons form the center of the atom, while the electrons whirl in orbit around the proton center, much like the moon orbits the earth.

If there were multiple earths very close together, and each earth had a moon, they could play a game of catch, passing the moons from one earth's orbit into the next earth's orbit. But the laws of our game require that each earth has one moon in orbit around it. If an earth didn't have a moon in orbit, having just passed it, it would have a positive charge on it, and be desiring to receive a moon. If an earth had an extra moon, having just caught one, it would have a negative charge to it, desiring to repel the extra moon. This passing of moons would continue until all the earths again had one moon in orbit.

receptors of electrons from other atoms and give the atom a positive charge that attracts negatively charged electrons from other atoms. This flow of electrons from one atom to the next is what we call "electricity."

The flow of electricity is more properly called "Current flow," or simply "Current." Current is the workhorse of electrical theory. The flow of electrons is what actually produces the power that we consume from the wall socket. "Voltage" is the pressure on the electrons to move from atom to atom. Voltage sits in the wall socket waiting to move electrons, but is only the "potential" for work. If you don't plug anything into the wall socket, Voltage can't do anything for you. But when you do plug something in, Voltage determines how fast those electrons will come streaming out, and consequently how much Current will flow; and how much power can be consumed.

CURRENT

The flow of electricity in a wire is like this passing of moons. A wire is made out a material called a conductor. A conductor is a material that likes to play this game of musical chairs. Wires are covered with materials called insulators, because insulators don't like to play this game. That keeps the passing electrons inside the wire.

When electrons are passed from an atom, they leave a "hole." Holes are potential

RESISTANCE

We are about to put the final piece of the puzzle together - resistance. Resistance "impedes" the flow of electrons. Resistance and Impedance are related, but for now we are only concerned with resistance.

Resistance is measured in ohms. A light bulb has resistance. The filament resists the flow of electricity, and glows white hot in doing so. This resistance causes the bulb to allow only a portion of the Current available in the wall socket, to flow out. A 100-Watt light bulb uses more electricity from the wall than a 60-Watt bulb, because the filament in a 100-Watt bulb has less resistance to it. This lower resistance allows more Current to flow through the bulb, consuming more power, producing more work, and making more light! This work, by the way is called "Wattage." If you multiply the Voltage times the Current the answer is in Watts.

OHMS LAW

Now comes the first hard part. You should study what I say here, and after it's had time to sink in, we'll continue.

Ohm's Law states the relationship between Voltage (E) measured in Volts, Current (I) measured in Amps, and Resistance (R) measured in ohms, as we said before. Ohm's Law is:

$$E = I \times R$$

Another way of saying this is the Current, multiplied by the Resistance, equals the Voltage. That is, the Voltage "dropped" across a Resistor is equal to the Current flowing through the Resistor. This is known as "The IR Drop." I'll remind you of this again later on.
From your high school math classes, I hope you remember, that you can switch this formula around, and get two more:

$$\frac{E}{I} = R \quad \text{and} \quad \frac{E}{R} = I$$

Just remember the Voltage (E) always goes on the top. With this simple formula you can do most of the math required in electronics. Study it; commit it to memory;

and think about the relationships:

1. For a fixed Voltage, as the resistance goes up, the Current goes down.

2. For a fixed Voltage, as the resistance goes down, the Current goes up.

3. For a fixed resistance, as the Voltage goes up, the Current goes up.

4. For a fixed resistance, as the Voltage goes down, the Current goes down.

5. To maintain a fixed Current, as the Voltage goes up, the resistance must also go up.

6. To maintain a fixed Current, as the Voltage goes down, the resistance must also go down.

Let's use our light bulb as an example. Remember earlier, you learned that the Wattage equals the Voltage times the Current? Wattage is a cousin of Ohm's Law. A short hand method of describing the combined terms of Voltage and Current. We'll use it now, since the only things we know about our light bulb is that the Voltage in the wall is 120 Volts, and that the manufacturer claims it to be a 100 Watt bulb. So to find the Current flowing through it (without setting up a real life test using meters) we divide the Wattage by the Voltage, and get:

$$\frac{100 \text{ Watts}}{120 \text{ Volts}} = .830 \text{ Amps}$$

Now to find the Resistance of the bulb, we can divide the Voltage by the Current and get:

$$\frac{120 \text{ Volts}}{.830 \text{ Amps}} = 144.5783 \text{ ohm}$$

25

Remember the IR drop? It's 120 Volts.

$$I \times R = E$$
$$or$$
$$.83 \times 144.5783 \text{ ohms} = 120 \text{Volts}$$

In other words, 120 Volts is "dropped" to zero (Ground) across a 144.5783 ohms resistor producing .83 Amps of Current flow, generating 100 Watts of electrical power (Wattage).

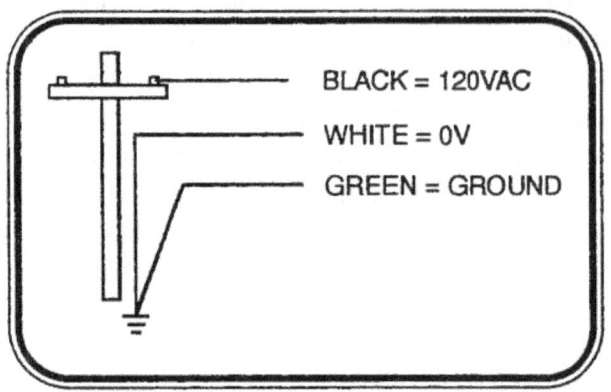

BLACK = 120VAC

WHITE = 0V

GREEN = GROUND

GROUND

Ground is a new term, so I'll define it. If you use a battery, you need to run two wires from it. One is labeled positive, and the other negative. Two wires are needed to develop an electrical potential, between them. To save running two wires to every house with electric service, the Electric Company uses the "Ground" itself as the "return," or negative wire. Only the positive wire is in the fuse box, with the negatively charged electrons actually being sucked out of the ground. Of course, the electric utility is also connected to the "Ground" ("Earth" in England). Ground is nominally zero Volts, although maintaining a good zero Volts ground is one of the problems facing electronic designers.

Now that we've covered the basics, we'll go on and look at the different types of electronic components; how they work; and which ones are the best to use in restoring vintage Fender amps.

CHAPTER 4
CAPACITORS

A capacitor consists of two metal plates separated by an insulator, so the two plates don't touch each other. One plate is Positive, and the other is Negative. This is true even in non-polarized caps. An older name for Capacitor is Condensor. In 1745, when the first Condensers were accidently brought into existence. They consisted of a glass jar, coated inside and outside with metal foil. It had a cork in the top, with a metal rod sticking through. A metal chain was attached to the end of the rod that was inside the jar, and this made the connection to the inside foil. The outside foil was, of course, available for conduction, outside.

When the German experimenters Ewald George von Kleist, and the Dutch physicist Pieter van Musschenbroek discharged these devices and found themselves stunned, they were "shocked," and even horrified. Von Kleist abandoned the experiments, and Van Musschenbroek proceeded with extreme caution. Van Musschenbroek did his experimentaion at the University of Leyden, in the Netherlands, and the Condenser became known as the "Leyden Jar." The original purpose of the Capacitor was to store electricity, predating the battery.

Earlier, in 1570, Gilbert studied the attractive forces produced by rubbing amber, diamond, sapphire, amethyst, opal, carbuncle, jet, and even ordinary rock crystal. He called these substances "electrics." If glass was rubbed with silk it gained a charge that would attract small pieces of other material. Since the charge "stayed" on the glass, it was named "Static" Electricity.

It was thought that metals couldn't be electrified. In 1729, an English electrician, Stephen Gray, electrified long glass tubes and found that corks placed into the ends of the tubes, as well as ivory balls stuck into the corks by long sticks, became electrified when the glass was rubbed. These experiments led him to believe that electricity was a fluid, and that this fluid could be transferred through a conductor. It was also discovered that metals can be statically charged, if care was taken not to conduct the charge away. Insulators were discovered that could hold a charge on its own little "Insula" or island. Using the properties of insulators and conductors, attempts were made to store electric fluid on larger surfaces, in order to build up a larger charge.

What was discovered is that a certain amount of negative electric charge could be put on a conductor plate. If a second positively charged plate was placed parallel on top of it, so the plates were close together, but not touching, a larger charge could be placed onto the bottom, negative plate. What happened was the top plate, which is positive, attracted the stored charge electrons to the top surface of the bottom negative plate. This meant that the bottom surface of the negative plate was empty, and now had room for more electrons to park. Being packed tighter, they are "Condensed," and the device has a higher "Capacity." The positive plate also has this higher capacity of charge.

It was further discovered that the increase in charge was affected by the space separating the plates. If it was a vacuum, or air, the plates would store about the same charge. But if a piece of glass was placed between the plates, the charge would increase 5 times. By placing a dielectric between the plates of a condenser multiplies the capacitance of the condenser by the dielectric constant.

Once charged, the cap could be discharged to create a spark, if you wanted to create a spark, that is. Think of this use as a "shunt" connection. "Shunt" is when the capacitor is connected "across" a Voltage, in parallel with the voltage source.

One plate is attached to the negative terminal of a 9 Volt battery, and the other plate is attached to the positive side of the battery. Electricity will flow from the positive battery terminal onto the "top" plate of the capacitor, until the plate almost reaches the same Voltage as the battery. If the battery is then disconnected, the capacitor will still have approximately 9 Volts stored on its plate, until it dissipates through the capacitor's losses.

The cap's insulator can be polystyrene, polypropylene, mylar, mica, paper, ceramic, or other things that are normally used to identify the type of cap, i.e. a "ceramic" disk cap, or a "silver mica" cap. Different construction techniques have different performance characteristics, and costs.

ELECTROLYTICS

Some caps, electrolytics and tantalums, for example, are polarized. They need to have one plate more positive than the other. They're labelled with a plus sign (+) for the Positive side, or an arrow pointing to the Negative. Electrolytic caps can be made smaller and cheaper than none polarized "film" capacitors of the same value, and are frequently used in the decoupling, or filtering role.

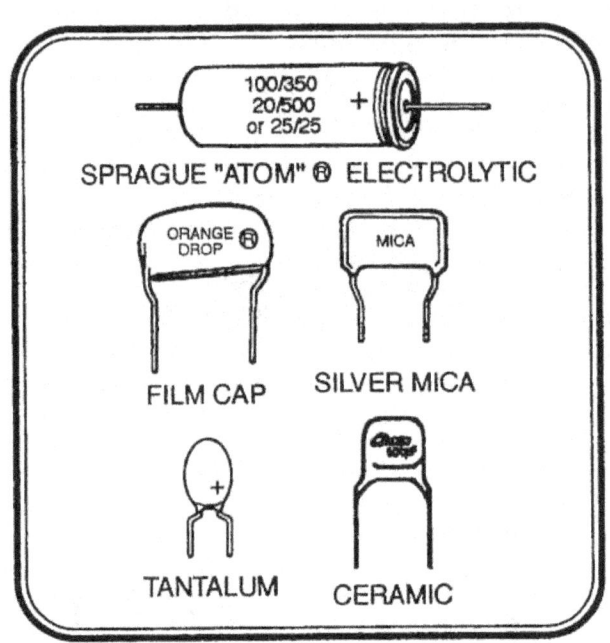

In the normal filtering application the Negative side is attached to Ground. In a bias supply, one that creates Negative Bias Voltages, the reverse is true. The Positive side of the cap is attached to Ground. The reason is, Ground is now more Positive than the Negative Voltage. The cap goes from -70 Volts on the Negative side of the cap, to 0 Volts (Ground) on the Positive side of the cap. The Positive terminal is more Positive than the Negative terminal, even though the Positive terminal is connected to Ground.

SERIES & SHUNT

The original use of capacitors was to store a charge of electricity, as a primitive battery. Once charged, it could be discharged to create a spark. Think of this as a "shunt" connection. Shunt occurs when the Capacitor is connected "across" a Voltage. That is, in parallel with it. One plate is attached to the Negative Terminal of a Battery, and the other plate is attached to the Positive side of the battery. Electricity will flow from the Positive battery terminal onto the "top" plate of the Capacitor, until the plate and the battery reach the same Voltage. If the battery is then disconnected, the Capacitor will still have 9 Volts stored on it, until it's discharged.

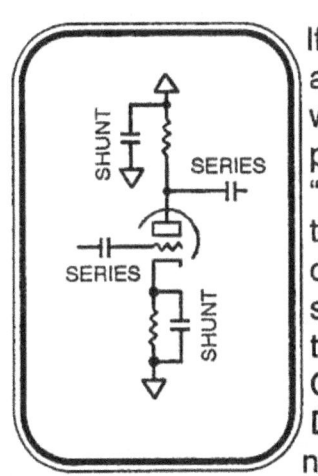

If an AC signal is applied to one plate whether the other plate is in "series" or "parallel" (shunt) with the Voltage, the AC component of the signal is transmitted through the Capacitor, but the DC component will not conduct.

FARADS

Think of the plates as a parking garage for electricity. How much electricity can park in the garage depends on the "capacity" of the cap. The capacity of a cap is defined in Farads, in honor of Michael Faraday. One Farad equals one coulomb per volt of charge on either plate. A Farad is quite large, so in practical terms the unit of capacity is the micro-Farad, or one-millionth of a farad. The abbreviation for micro-Farad is uF. The u stands for a script style of small m, not to be confused with mF, which would mean

milli-Farad, or one-thousandth of a Farad (which is never used).

Continuing on to even smaller values is the nano-Farad, and then the pico-Farad, or pF, commonly pronounced "puff," which equals one trillionth of a Farad. A 250 puff cap then is .000,000,000,250 Farads. On some older Fender schematics, MMF was used as an abbreviation for pico-Farads. It stands for micro-micro-Farad.

On an English schematic you might see 2N2. This translates into 2.2 nanoFarads, or .000,000,002,200 Farads. Just look at the N as a comma, add zeros to form three decimal places to the right, and read it as 2,200 pF, or three decimal places to the left for .0022 uF. Easy enough?

1 F = One Farad or 1F

.001 F = One milli-Farad or 1mF

.000,001 F = One micro-Farad or 1uF

.000,000,001 F = One nano-Farad or 1nF or 1,000pF

.000,000,000,001F = One pico-Farad or 1pF (puff)

As a practical matter, most caps are denominated in micro-Farads, and all caps are rated for Voltage, which is how high of a Voltage can be applied to the two plates before the insulator fails.

FILTER CAPS

After the rectifier in a power supply circuit, a capacitor is connected, in "shunt" to "filter" (filter cap, get it?) the AC wall electricity into DC. Remember that as the

29

Voltage goes up, the cap stores electricity on its plates. When the Voltage goes down, it releases that electricity, creating an "average" Voltage, and smoothing out the AC vibrations. This "filters" the AC electricity, and turns it into DC electricity. This is what a power supply does. It takes the AC from the wall, rectifies it, filters out the DC "pulses," and leaves "pure" DC.

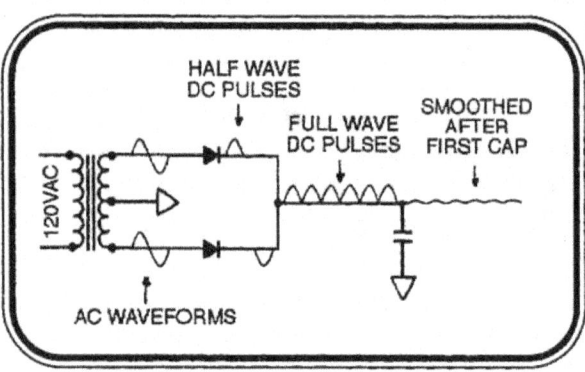

When we talk about "filter" caps, we generally mean electrolytics. They are cheaper, and smaller than other types, with the only disadvantages being that they have a positive and negative terminal, and have to be hooked up properly. They are also relatively "slow" in discharging their stored electricity into a circuit. This is fine in a power supply circuit which operates at the realtvely slow 60-120 cycles.

Electrolytics should last between 10 and 20 years in a properly designed amp, so you know a lot of them are due to be replaced. When they fail they will either lose their capacity and stop filtering out the AC from the DC, causing 60 cycle hum. For technical reasons, most power supplies operate at 120 cycles, but to most ears, it still sounds like 60 cycle hum.

Sometimes the filter caps will short out, drawing enough Current to blow a fuse, or just sizzle as the electrolyte boils from the

heat generated, and vibrate the choke and/or transformer as the amp tries to blowup or burn up. God, I love electronics.

MOTORBOATING

Another defect from a failing filter cap, or an improperly designed amp, is "motorboating." With inadequate filtering, some of the audio signal going through the amp gets superimposed onto the DC supply and bleeds into previous stages of an amp, causing low frequency feedback, causing the amp to oscillate at a low frequencies. This noise sounds either like a motorboat, or a '55 Mercury running without a muffler.

TESTING CAPS

If you have a capacitance meter, you can test the capacity of your caps (see the WARNING about discharging caps). The problem with meters is they run at low voltages, and they don't show how the cap is performing under the high voltages of a vacuum tube amp.
Or, you can measure the amount of AC on the primary filter stage. It should be between 1% and 2% of the DC value. That is, a supply of 400 Volts DC, can have between 4 and 8 Volts of AC on it without any concern.

Or you can do a visual check. If you see liquid, or a white powder, or a stain on the cap from a leak; or if you see a bulge at the positive (+) side of the cap, or the pressure relief valve is popped, or protruding from the positive end; you can bet the end is near. Replace with only quality American Made caps for good tone, long life, and reliability.

WARNING — DANGER!

Never work with power supply caps unless you're sure they're discharged. The best way to do this is to have some type of discharging resistor permanently added to the circuit. A 220K, 1 watt resistor connected across each of the 70/350V caps at the first filter stage will slowly bleed off the high voltage stored in the supply. These resistors are quite often found in Fender amps, underneath these two large caps. The reason they were added was to balance the Voltage Drop between the two caps that are stacked in series.

An easy way to discharge the caps is to warm the power tubes up. Then, when turning off the amp, leave the Standby switch "On." This should discharge the caps within a few seconds. Check the voltage at the first stage of the power supply, where the voltage is the highest. A probe to discharge caps can be made by attaching a wire to a 100K to 470K, 1 watt or larger resistor, and touching or attaching one end of the wire to the amp chassis, and touching the other end to the high voltage point. This is extremely dangerous if you accidentally touch the wire. My soldering iron has a grounded tip, which will discharge filter caps when touched to the hot lead. This is a nice safety extra, if there's a little bit of voltage left in the cap..

CREATING OTHER VALUES

If you need a larger cap value, you can connect two caps in parallel, and their capacitance adds together, at the same voltage. If you need a smaller cap value, you can put two caps in series, which would cut the capacity in half, but would double the Voltage rating.

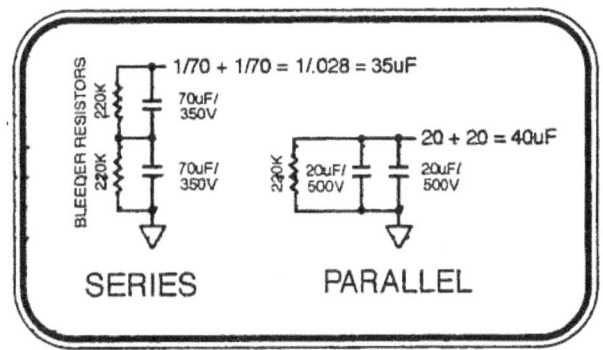

In a lot of Fender amps, you have two 70uF caps, rated at 350 Volts each, connected in series. This results in a 35uF filter, rated at 700 Volts.

In some amps you might find 2x50uF caps rated at 500 Volts each, connected in parallel. This results in 100uF filter at 500 Volts.

USING LARGER CAPS

When replacing caps, you might not be able to find the exact value you're looking for. Older Fender amps used 16uF cap. With lower demand, the number of different cap values being manufactured has been reduced. The standard replacement value for 16uF, is now 20uF. If you use the larger value cap, you'll get better response from the amp.

If you use a cap with a higher Voltage rating, you'll get longer life out of the cap, except that electrolytics work best when used close to their Voltage rating. If you have to replace a 70uF at 350 Volts, and you have to choose between a 100uF at 350 Volts, or an 80uF at 450 Volts, I would use the 100uF at 350 Volts, for this reason. In some Fender amps, there are two 70uF, 350 Volt caps in series, providing a 35uF filter at 700 Volts. Since the amps rarely exceed 550 Volts, you have plenty of Voltage rating headroom. Two 80uF caps would provide a 40uF filter. The two 100uF caps would provide a better 50uF filter.

CAUTION: Be warned against using too large of a cap to filter the Bias supply. I would not exceed 100uF under any conditions. The size of the Bias Supply Cap affects the Time Constant of the Bias Supply, adversely affecting the tubes ability to respond to varying voltage conditions. An example would be the long time it takes to charge a large Bias Cap when the amp is first turned on. For those seconds the amp is basically running without bias. This can damage your power tubes, especially if they're already warmed up and the amp is turned off, and then back on again, within a few minutes. The power tubes only need 1 or 2 milliamps of bias current, which would take a long time to change the charge on a large cap.

Another example of this causing a time constant problem is the changing of operating points due to fluctuations in the AC wall voltage. If your Plate Voltage goes up you want your bias to follow upwards (more negative). Sometime today would be nice. If you really think you hear "a dramatic improvement in tone" from a large bias cap, and you want the effect of a larger cap without the time constant hassles, then change the 220K bias resistors to 100K, a mod I'd recommend anyway. You'll get the effect of a larger cap without the time constant problems. This is another example of bad information being out there, written by guys who desperately need SOME kind of a "secret!"

IMPEDANCE

The electricity that comes out of the wall is called 117 Volts "AC." The AC stands for "Alternating Current." The Current (flow of electricity) alternates back and forth, because the Voltage is fluctuating above, and below ground. At one time sending Current to the toaster, and on the next half cycle, pulling it back. In this book, we'll loosely use AC to mean any voltage that has a flucuating frequency, whether it remains above ground; below ground; or actually reverses itself across ground.

The Voltage is the "push" that causes electricity to flow through a resistance. Resistance is what impedes the flow of electricity. As the Voltage goes up, more electricity (Current) gets pushed through a fixed resistance. If the resistance goes up, then less Current will flow for a given Voltage.

There are three types of resistance, all denominated in ohms. Straight resistance doesn't change if the Voltage applied to it is fluctuating. The signal from a guitar is a fluctuating Voltage. The other two types of resistance change with the frequency of the signal. The term "impedance" is used to describe a resistance that changes with frequency. Speakers are rated in ohms of impedance, but many people don't realize that a speaker's impedance is a nominal rating, and the actual impedance can drop dangerously low at certain frequencies, stressing a poorly designed amplifier.

IMPEDANCE OF CAPS AND COILS

A capacitor's Impedance goes down as

the frequency goes up. Capacitors like to pass high frequencies. A coil of wire's Impedance goes down as the frequency goes down. Coils like to pass low frequencies. You can guess then, that by using resistors, caps, and coils, you can design a circuit that controls how much, of what frequencies, can get through a circuit. This is the basis of tone controls. Inductors are rarely used in guitar amps, due to their high cost. The graphic EQ on Fender amps is an exception. Guitar amps generally use only resistors and capacitors to accomplish their tone control.

Voltage with a frequency of zero is called DC. Batteries put out DC. DC can have an AC Voltage super-imposed on top of it, such that the Voltage might vary from 9 Volts up to 10 Volts, and then down to 8 Volts, and back to 9 Volts. The DC component is 9 Volts, and the AC component is 2 Volts. This is exactly what bias is. A 2 volt AC signal, being offset by a 9 volt DC voltage, in this case.

Since capacitors consists of two conductive plates that don't actually touch each other, they have the unique property of being able to completely block the flow of DC electricity through them, while at the same time passing an AC Voltage through them, due to their "capacitive effect."

HOW AN AMPLIFIER USES CAPACITORS

Now, the way an amplifier works, is it takes the 60 cycle AC power from the wall, and, with a power supply circuit, takes the shake out of it, turning it into a large DC Voltage. Then a small AC guitar signal shakes the large DC Voltage, turn-

ing it back into a large AC Voltage, which drives the speaker.

As part of this process, the AC signal Voltage needs to be separated from the DC power supply Voltage. Because capacitors block DC, while passing AC, they are used to separate the two. In this application they are called "coupling" capacitors, because they "couple" the AC signal from one amplifier stage to the next.

In a vacuum tube amp, capacitors are usually not used to couple the power tubes to the speakers, because an output transformer is needed anyway, and it accomplishes the same DC isolation. Since capacitors resist the flow of low frequencies, totally stopping a frequency of zero, you need bigger, and bigger capacitors, to pass lower, and lower frequencies.

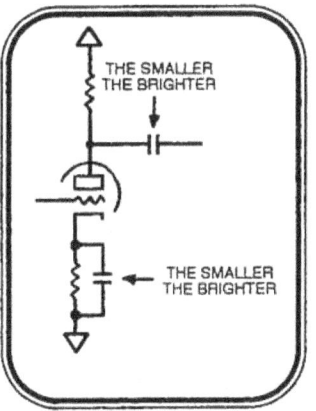

So, a larger cap between two amp stages, allows more low frequencies to be passed through. This trick can be used to equalize an amp's frequency response.

In high fidelity amps, the coupling caps are chosen to pass very low frequencies. This can muddy up a guitar amp's tone. To remove that mud, coupling caps are made smaller. This is especially helpful in distortion circuits, where the low frequencies can modulate distortion notes, and chords, causing them to "roll." Using a larger cap between stages passes more low end, and a smaller one passes less.

Don't get the idea that I'm saying you can improve the low end response of an amp by just increasing the coupling cap size. There are many other factors that affect the sound of an amp, and I can't just give you a value to use. The value of the cap changes with the total impedance of the circuit, the transformer, the tubes, what part of the circuit we're talking about, and what tone you're looking for. I'll discuss this more in a future article on distortion.

TROUBLESHOOTING COUPLING CAPS

The output Plate of a tube has a DC power supply Voltage on it of up to 330 Volts, and an AC signal Voltage of up to 70 Volts on it. The coupling cap connects the output Plate of one tube to the input grid of the next tube. Its purpose is to let the AC signal Voltage through, while blocking the DC supply Voltage.

The input grid of the next preamp tube is normally tied to ground, and has zero DC Volts on it. This zero Volts Grid is part of the biasing scheme. When a coupling cap fails, it usually starts to let DC leak through. This DC Voltage gets on the input grid of the next tube stage, and wreaks havoc with the biasing scheme; causing fluctuations in the gain; a loud crackling noise that makes the amp unlistenable; a change in gain that continues after you stop turning the pot; or, if you have a scratchy popping sound on your controls, that doesn't go away after you clean them (except "Presence" controls which are naturally scratchy due to the DC on them, which is part of the design). If you have these problems it indicates that your coupling caps could be leaking DC through, and need to be replaced.

VIBRATO

Caps are also used to time the oscillator in the vibrato circuit. If the vibrato doesn't work, or the speed doesn't seem as fast as it used to be, suspect the caps.

OLD AMPS

In an old amp, if you need to change one cap, you should change them all, unless you intend to keep the amp in as original condition as possible. I personally believe that if you don't change all the caps, you'll never know what the amp sounded like when it was new. Also, if you only change the bad cap, the amp will just keep coming back for one cap change at a time. Meanwhile the amp will sound weak, and it won't be reliable.

If you have a capacitance meter, you can test the capacity of your caps, but meters run at low Voltages, and they don't show how the caps are performing under the high Voltages of a tube amp. Don't trust a meter when testing high Voltage coupling caps.

DECOUPLING AND SUPPRESSION CAPS

Large caps pass more low frequencies.

A Voltage with a frequency of zero is called DC. Batteries put out DC Voltage. If the Voltage fluctuates up and down at any frequency, it's called AC. Because a capacitor consists of two plates that don't actually touch each other, they have the unique property of being able to completely block the flow of DC electricity through them, while at the same time passing AC.

DECOUPLING CAPS

Decoupling caps are used to detach (decouple, bypass) one part of a circuit from another. This is the opposite of the coupling cap story. Remember, a DC Power Supply Voltage remains fixed at a certain level, and is used to setup operating points (bias) in a circuit. Decoupling caps are used to stop the AC signal Voltage from affecting the DC operating Voltages in an amp. That's the complement to coupling caps keeping DC Voltages from leaking through and affecting the next DC operating point in an amp. If an AC signal comes through the circuit, it can add, and subtract from the DC operating point, and cause problems. The decoupling cap is used to provide an AC path to ground, separating it from the DC Voltage, and helping to maintain the DC operating point at a steady level.

Remember, even though the cap is connected to ground, DC Voltage can't pass through it, allowing the cap to short out the AC Voltage, while not affecting the DC Voltage.

Where do you find decoupling caps in an amp? They're what you most commonly see as Cathode caps. You know, the little 25uF at 25 Volts Electrolytic Caps found on the circuit boards of Fender and Marshall amps. These caps are connected to pins three, and eight on the preamp tubes. To understand how they work, we must discuss how the preamp tubes operate.

DECOUPLING PREAMP TUBES

A vacuum tube is nothing more than a light bulb with a "Plate" inside it. The Plate was inserted, and a positive Voltage applied, as a way to attract the black soot generated as early bulbs burned. It was noticed that a flow of Current was created between the filament and the positive Plate. This was the invention of the "diode." It conducts Current in only one direction. The way it works, is that the filament (which in this case is also the Cathode) emits electrons (negative charge) that are attracted by the positive charge on the Plate (anode). The amount of Current flow established is based on the temperature of the Cathode, and the amount of positive Voltage on the Plate. Without a way to control this flow of Current, the diode runs full on all the time.

To control this flow of Current, a third element (the control grid, sometimes labeled G1) was added between the Cathode and the Plate. This grid interferes with the flow of Current by electrically hiding the Plate from the Cathode. It does this by being negative relative to the Cathode. This negative Voltage is called "bias." Now, this bias Voltage doesn't have to actually be negative. It just has to have a lower Voltage than the Cathode. One way to do this is to make the Cathode more positive. In a preamp tube, this positive Voltage is created by the Cathode resistor. To understand this, we're going to have to discuss Ohm's Law.

OHM'S LAW

As we saw earlier, the most basic formula in electronics is Ohm's Law. Simply put, it is IR=E. The "I" stands for Current. The "R" stands for resistance, and, the "E" stands for Voltage. To make this as simple as possible, think in terms of water. The Voltage is the water pressure; the Current is the quantity of water that flows through a pipe; and the resistance is the size of the pipe. If you make the pipe bigger (lower resistance), more water will flow (Current) for a given pressure (Voltage). If you increase the pressure, more water will flow through a small pipe. If you multiply the amount of water flowing by the resistance of the pipe, you'll calculate the pressure needed to create that flow. In electronics, this is called the "IR drop." For a given amount of Current flow through a given resistance, a certain amount of Voltage "drop" (loss) is created. If you connect a resistor between a 9 Volt battery and ground, there would be 9 Volts at the top of the resistor, and zero Volts (ground) at the bottom.

CATHODE BIAS

If you place a resistor from the Cathode of a tube to ground, the Current flow through the tube must also come up through the resistor. Does it make sense to you that if 300 Volts is on the Plate of the tube, and the Cathode resistor is connected to ground, then the 300 Volts must be "dropped" across the entire tube/resistor stack? 300 Volts at the top, and zero Volts (ground) at the bottom. Since all the Current flows through both the tube, and the resistor, the portion of the Voltage "dropped" across the resistor equals the resistance of the Cathode resistor multiplied by the amount of Current flowing through it. This is an "IR drop." The result is that the top of the Cathode resistor, which is connected to the Cathode, is lifted to a Voltage higher than ground. If the Cathode is higher than ground, and the grid is at ground, then the grid is negative relative to the Cathode. This is how "Cathode Self-Bias" is generated.

CATHODE DECOUPLING CAP

When a signal is placed on the grid, it makes the grid more positive, turning the tube on, and increasing the Current flow through the tube, and the Cathode resistor. This increases the "IR drop" across the Cathode resistor, raising the Voltage on the Cathode. This results in the Cathode Voltage following the signal Voltage up, which reduces the potential Voltage difference between the Cathode and the grid. This reduces the amount of gain you can get out of the tube. This coupling between the Cathode and the grid is what a decoupling cap takes care of. When the Cathode fluctuates with the signal changes, these changes are AC and can be bled off to ground with a Cathode decoupling cap. This helps to maintain the Voltage difference between the Cathode and grid, allowing greater dynamic range, or gain, out of the tube.

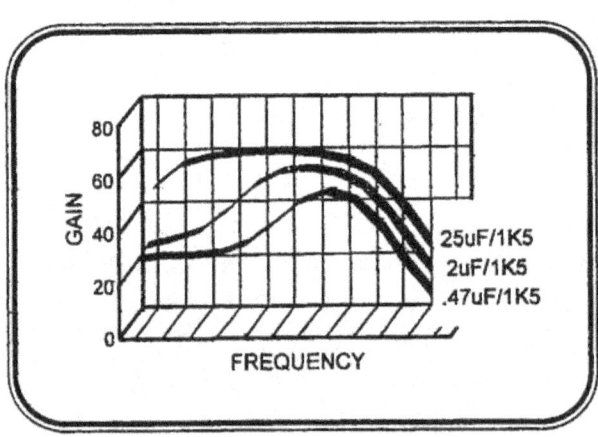

Now, for the trick. Since a larger cap passes lower frequencies, the frequency response of a tube stage can be adjusted by changing the size of the decoupling cap (see figure 1). You might notice that a Fender clean channel has a 25uF cap, while a distortion amp stage might have a .47uF cap. The difference is that the smaller cap doesn't decouple the low frequencies, allowing the Cathode to fluctuate at these frequencies, reducing the low frequency gain. This tends to strip the low frequencies out of the distortion circuit, reducing the muddy, rolling sound that results when the harmonics generated by those clipped low notes beat against the mid-range fundamentals.

Some Marshall amps have one channel with a 25uF cap on the first stage, and the other channel has a 320uF cap. The 320uF cap is the "bass" channel. If this channel sounds too muddy, reduce the size of this cap until you're happy. For most people, 25uF is all the low end needed. Anything bigger, and you'll be turning down your bass control. The interesting part of this, is that without the cap, the frequency response is actually flatter, although the gain is reduced in half.

PREAMP GAIN

While we're on the subject, let's discuss the gain of these preamp stages. The "best" resistor value for the first stage of a preamp is 820 ohms. The reason for this, is that 820 ohms biases the first stage to receive the largest input signal without asymetrical clipping. An 820 ohm resistor stage with no cap on it will take almost 1 Volt of input signal without clipping. Are the steel guitar players listening? As the resistor value goes up, the gain of the stage is reduced. When you reach 47K, you're basically at unity gain (1 Volt in, 1 Volt out).

Guitar preamp circuits are generally this self-biased and bypassed setup, and only require service when the "Cathode cap" or "Cathode resistor" fails. This will generally

be noticed by a loss in gain. As the Cathode cap ages it loses its capacity. Resistors generally age up in value, again reducing the gain, unless they actually burn open, in which case the tube stage will stop working altogether.

CLASS-A POWER

The power tubes in a Class-A power amp are also generally self-biased, and bypassed. This cap usually fails by shorting out, destroying the bias, and gain, and possibly smoking, and burning. Fender tweed amps, many Gibson amps, and Vox amps are the most common Class-A power amp designs. I would suggest replacing the 25 Volt decoupling cap on Fender Tweeds with a 50 Volt rating. The reason for this is that the higher Voltage rating cap will age slower, and be less susceptible to over Voltage destruction, which can happen with these amps.

You can use a capacitance meter to test the capacity of your decoupling caps, and unlike filter, and coupling caps, which run at high Voltages, decoupling caps can be accurately tested with a low Voltage cap meter. You'll most likely find a 25uF cap that reads low, or even zero. If the cap reads less than its rating capacity, that is 22uF for a 25uF cap, replace it. A lot of imported caps fail this test even when brand new. This is another reason I stay away from them.

HIGH VOLTAGE DECOUPLING

The Power Filter Caps we discussed earlier, are also a type of Decoupling Cap. These caps remove the effects of the AC signal as the tubes consume power out of the DC supply. If these caps fail, or are inadequately sized, the amp will motor-boat, in a low frequency oscillation.

Small caps are also sometimes used for power supply bypass close to integrated circuits. These provide a small, fast amount of power close to the consumer of the power, decoupling the AC signal from the DC supply.

SUPPRESSION CAPS

The other use of caps, is for supression. These provide a path for high frequency oscillations to dissipate. You might find a small cap across the Plates of the driver/inverter tube, or a cap across one of the Plate Resistors in a distortion stage, or the famous 2000pF caps on the power tube grids of Silver-face Fenders.

Many Tweed amps had suppression caps across the inverter Plates. These caps are 47pF to 100pF. The best size for a supression cap is as small as possible, as long as it still suppresses the oscillations. Or, as large as possible, as long as you can't hear it. Test your hearing by changing to higher and higher values until you hear a loss of high frequencies. Then go one step smaller, and that's the maximum

size suppression cap. If you still have oscillation you'll have to find another way to fix it.

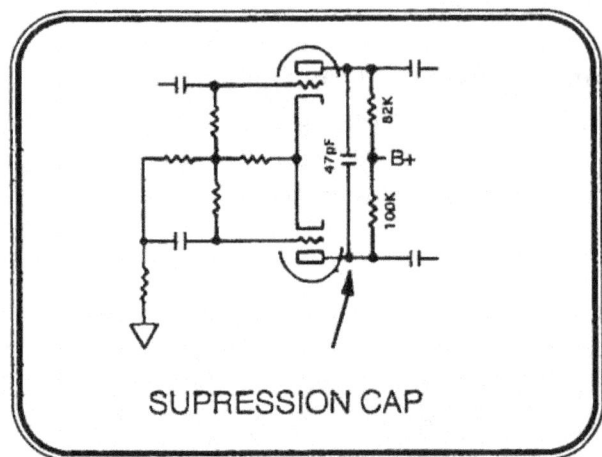

SUPRESSION CAP

Let me explain that last statement. An amp oscillates either because it has phase shift at certain frequencies that turns negative feedback into positive feed-back; or it's output signal is coupled to its input by an antennae effect. Any of these problems are defects. They should be fixed by the designer, and phase shift at high frequencies is easily fixed by sup-pressing those frequencies, if they are inaudible. Remember, the highest perfor-mance aircraft are the ones that are unstable. The reason they can turn so fast, is that they don't like to fly straight.

CHAPTER 5
RESISTORS

A Resistor has a fixed Impedance that doesn't vary with frequency. Resistors are used to divide Voltage, limit Current (Current limiting resistors), or to create Voltage drops (IR drops).

Resistors conduct electricity. How well, or rather how poorly, they conduct electricity is measured in a unit called Ohms. If you make the resistor out of different materials; or make it longer, shorter, thicker, or thinner; you can get a higher, or lower resistance. This resistance is used to do the following things.

CURRENT LIMITING

Resistors can be used to limit how much electricity flows through a circuit. These are called Current Limiting Resistors. Examples of this are the resistors used to limit the Current flow through an LED. With a low resistance, more Current will flow. With a high resistance, less Current will flow. If the resistance stays fixed, and the Voltage is raised, then more Current will flow for this higher Voltage. This could be solved by changing to a yet higher value Resistor to further limit the now larger Current flow. It's a matter of balance.

Using a light bulb in series with an amp under test, is an example of Current limiting. If the amp shorts out, there will still be a light bulb in the circuit, limiting how much Current flows. After all, a shorted amp with a light bulb attached is nothing more than a light bulb with a switch

attached, and the switch is "on."
A larger light bulb has less filament resistance, allowing more Current to flow through it, resulting in a higher Wattage bulb (Voltage times Current equals Watts), which generates more light.

Think of Current as power; Voltage as the potential to do power; and resistance as the device that controls the application of the power. Voltage without Current is power waiting to happen. Voltage without resistance is a short circuit. Current without Voltage doesn't exist. These three elements are what electricity is all about, and is expressed in Ohm's Law, known as $I \times R = E$.

If you attach a tungsten street light bulb to your wall, it will probably blow your fuse, because the bulb's resistance is so low, it will allow 20 Amps of Current to flow through it. The fuse sees the bulb's low resistance as a short circuit. Make the resistance high enough, and it will limit the Current through the fuse (the opposite of a short). The Current draw could be cut in half by putting two street light bulbs in series. They would now be operating on half the voltage divided between them, and they would not burn as bright.

VOLTAGE DIVIDING

Resistors can also be used to divide Voltage. If you connect two resistors of equal value, end to end (in series) across a 9 Volt battery, the center point will have 4.5 Volts on it. This is how potentiometers (pots) work. They use a "wiper" to divide one long resistor into two pieces, and can adjust the dividing point between them, by moving the wiper.

The Voltage "dropped" across each resistor depends on the position of the wiper. Voltage drop is a common electrical term, also called the "IR Drop," and something that you'll find very useful, once you learn how to use it.

"IR DROP" or OHMS LAW

The "IR Drop" formula is the definition of Ohm's Law, the most basic formula in all of electricity.

$$I \times R = E \quad \text{IS OHM'S LAW}$$

If you multiply the Current (I) times the Resistance (R), it equals the Voltage dropped (lost? dissipated?) acoss a resistance.

In other words, if you connect a 200 ohm resistor across a nine Volt battery, one end will have nine Volts on it, and the other end will have zero Volts on it. This seems obvious.

Can you figure how much Current is flowing from the battery, through the resistor? From your high school math classes, you should remember, that if you switch our Ohm's Law formula around you get two more:

$$I \times R = E \quad \text{inverted equals}$$

$$E / I = R \quad \text{and} \quad E / R = I$$

Just remember the Voltage (E) always goes on the top. With these three simple formulas you can do most of the math

required in electronics.

Since we know the Voltage, and resistance, we use E/R=I, or 9 Volts divided by 200 ohms to solve for .045 Amps (45 milliAmps).

The next formula we need to know is how to compute power (measured in Watts).

$$I \quad x \quad E \quad = \quad P$$
$$.045 \text{ Amps x 9 Volts } = .405 \text{ Watts}$$

So, our 200 ohm resistor across our nine Volt battery needs to be at least a 1/2 Watt resistor. If you use a 1/4 Watt, it will burn up. If you use a 1/2 Watt, it will run hot, but survive. If you use a 1 Watt, it won't get much more than warm.

Now, let's figure a preamp Plate Resistor. The tube book specifies a maximum of 330 Volts on the Plate of a 12AX7. A common value for the Plate Resistor is 100,000 ohms. So, 330 Volts divided by 100,000 ohms equals .0033 Amps. Then, .0033 Amps times 330 Volts equals 1.089 Watts. So, if the preamp tube shorts to ground (very rare) and all the Plate Voltage is dropped across the Plate Resistor, its maximum dissipation would be 1.089 Watts.

If you use a 1/2 Watt resistor in this position, and have a catastrophic failure of the preamp tube, the resistor will last for awhile, but eventually blow. This can be thought of as a safety feature, acting much like a fuse. If you notice the amp not working, and replace the defective tube, everything will be fine. If you don't replace the tube, the resistor will eventually burn open, and force you to have the amp fixed. Don't get paranoid about this. It is extremely rare.

If a coupling cap shorts out (more likely, and still extremely rare), it is usually followed by at least another 100,000 ohms to ground, cutting the Plate Resistor's dissipation to .272 Watts.

330 Volts / 200K ohms = .00165 Amps
.00165 Amps * 330 Volts = .544 Watts
.544 Watts / 2 resistors = .272 Watts each
Fender used 1/2 Watt resistors in this position, and after forty years almost all the failures I've seen are due to noise, or a shorted tube.

By the way, in the 1960's, 1/2 Watt resistors were cheaper than 1/4 Watt. Today, it's the opposite. You would think that the 1/4 Watts would always have been cheaper, because they're smaller - using less material. But, the cheapest parts, are the ones that have the highest production quantities. 1/4 Watt resistors became more popular as transistors started a miniaturization craze. A premium was paid for them, in order to reduce the size of products. Early transistor products used 1/2 Watt resistors because they were cheaper (look inside your old foot pedals). This is self-driven, because the cheapest parts are the most in demand, further increasing production, and lowering price, putting them back in demand. More important than overkill on Wattage, is the type of resistor used. There are five basic types of resistors; Carbon Composition; Carbon Film; Metal Film; Metal Oxide; and Wirewound. Let's review them and their differences.

CARBON COMPOSITION RESISTORS

Carbon Composition resistors are the noisiest, and are intended for general purpose use. They are the ones you've seen the most of in vintage amps. They're made by taking a measured amount of carbon paste, mixed with impurities to change its resistance, and pressure molded into a block. The quantity of carbon, the impurities, the cross sectional area, the length, and the pressure of molding, all determine the part's eventual resistance. An advantage in having a lot of material mass is its ability to absorb current surges found in tube amps. But there's another reason.

These resistors sound great in vintage amps. Care should be taken when replacing these resistors. They make a difference in certain key positions, and no difference in others. Areas of concern would be the 68K input jack resistors, the 220K isolation and channel mix resistors, and the infamous 100K plate resistors. These resistors should remain "Carbon Comp."

The problem with these resistors is something called thermal noise. When a material is heated, it emits Thermal Noise due to the random bumping of electrons, moving around as they become energized by heat. These electrons add to the signal, causing the hiss we've all heard. Using a 1 Watt carbon composition resistor, while reducing the resistor's temperature and thermal noise, increases the amount of carbon material that's emitting the noise. Another problem with carbon comps, is the uniformity of the paste mix. If you get a hot spot in the mix, it will become noisy. Static, hiss, or a "popcorn" noise are the result.

Sometimes a resistor will snap, crackle, and pop because of a broken connection between the wire lead, and the carbon block. This noise is caused by little sparks jumping between the lead and the resistance block.

TOLERANCE

Carbon comps are produced by molding a large quantity of resistors, and then measuring them to determine their value. If a 1,000 ohm resistor is the target, and some come out as 900 ohms, they can be marked as either 1,000 ohm, or 820 ohm, both with a 10% tolerance.

Carbon Composition resistors are the choice for vintage amp restoration. They are quite expensive these days costing around $1 each, but do affect the tone.

RESISTOR COLOR CODE

Most resistors are labeled with color coded stripes. It's something you really should memorize. It is:

VALUE	TOLERANCE
BLACK = 0	NO BAND = 20%
BROWN = 1	SILVER = 10%
RED = 2	GOLD = 5%
ORANGE = 3	RED = 2%
YELLOW = 4	
GREEN = 5	
BLUE = 6	
VIOLET = 7	
GRAY = 8	
WHITE = 9	

Resistors have three, four, or five stripes on them. The first, and second stripes (closest to one end) are the first two value numbers. The third stripe is the number of zeros that follow. So, Brown, Black, and Yellow would be 1 0 0000, or 100,000 ohms. Yellow, Violet, Orange would equal

4 7 000, or 47,000 ohms. Thousands of ohms are often abbreviated "K." So, our resistors could be called 100K, and 47K ohms, respectively. One million ohms is abbreviated "M," or megohms (megs). Orange, Orange, Green equals 3 3 00000 ohms, or 3,300,000 ohms, or 3.3M ohms. The English use a system, which I've adopted, placing the M or K where the decimal point usually goes. So, instead of writing 3.3M, you would write 3M3. This avoids the confusion on schematic copies, when the decimal points disappear.

The fourth color stripe on the resistor is the tolerance, and the fifth stripe (if there is one) is the reliability factor. Resistors come in values by the "tens," or "decades." The stock, and available decade values for common resistors are:

If you want a more precise value, you'll have to look at 1% metal film resistors. We'll discuss them in a minute.

CARBON FILM RESISTORS

The carbon film resistor is less noisy, and cheaper than the carbon composition. It's formed by breaking down hydrocarbon gas at high temperatures, forming a deposit of carbon film on the surface of a ceramic cylinder. Final resistance is obtained by cutting a spiral groove in the film with a diamond wheel to form a long spiral of resistance film wrapped around the ceramic core. The resistor is then painted with a high Voltage insulator to protect the film from humidity, and insulate the part from shorting out other parts it might touch. Because of the small amount of carbon film material, they're more susceptible to overload breakdown. Their popularity is due to their low price.

RESISTOR VALUE CHART

STOCK VALUE	1	10	100	1K	10K	100K	1M	10M
AVAILABLE	1.1	11	110	1K1	11K	110K	1M1	11M
STOCK VALUE	1.2	12	120	1K2	12K	120K	1M2	12M
AVAILABLE	1.3	13	130	1K3	13K	130K	1M3	13M
STOCK VALUE	1.5	15	150	1K5	15K	150K	1M5	15M
AVAILABLE	1.6	16	160	1K6	16K	160K	1M6	16M
STOCK VALUE	1.8	18	180	1K8	18K	180K	1M8	18M
AVAILABLE	2	20	200	2K	20K	200K	2M	20M
STOCK VALUE	2.2	22	220	2K2	22K	220K	2M2	22M
AVAILABLE	2.4	24	240	2K4	24K	240K	2M4	
STOCK VALUE	2.7	27	270	2K7	27K	270K	2M7	
AVAILABLE	3	30	300	3K	30K	300K	3M	
STOCK VALUE	3.3	33	330	3K3	33K	330K	3M3	
AVAILABLE	3.6	36	360	3K6	36K	360K	3M6	
STOCK VALUE	3.9	39	390	3K9	39K	390K	3M9	
AVAILABLE	4.3	43	430	4K3	43K	430K	4M3	
STOCK VALUE	4.7	47	470	4K7	47K	470K	4M7	
AVAILABLE	5.1	51	510	5K1	51K	510K	5M1	
STOCK VALUE	5.6	56	560	5K6	56K	560K	5M6	
AVAILABLE	6.2	62	620	6K2	62K	620K	6M2	
STOCK VALUE	6.8	68	680	6K8	68K	680K	6M8	
AVAILABLE	7.5	75	750	7K5	75K	750K	7M5	
STOCK VALUE	8.2	82	820	8K2	82K	820K	8M2	

METAL FILM RESISTORS

Metal film is the most reliable, quietest, and most expensive resistor. Its made by depositing a film of nickel-chromium onto a polished high alumina ceramic cylinder, and then cutting a spiral groove around it with a diamond wheel, or a laser beam. Although they have a small amount of noise producing "mass," you can imagine that a film of nickel and chrome, is very tough. They are not available in values over 1 meg.

Most people think of metal film as "those one percent resistors." Actually, one percent is sloppy for metal film. They are available in 1%, .1%, and .01% tolerance. I use metal films for their tight tolerance, low noise, and reliability. The larger the value of the resistance, the more noise voltage is created. Metal films make good candidates for the 1M resistors on the input jacks, but not the 68K's, which should be Carbon Comp. Don't bother changing yours though, unless you're having trouble with the original ones.

MILITARY PARTS

Metal film resistors are frequently referred to as "mil standard" parts. So, when companies advertise "military grade parts," there's an implication that they're using metal film resistors. Actually, even 1/2 Watt carbon composition resistors are covered in the military standard as type RC20. Metal films are called RN60 for 1/4 watt military rating, which is a 1/2 Watt commercial rating, and RN65's for 1/2 Watt military, or 1 Watt commercial. The RNxx number is followed by a letter which represents the thermal drift rating. The smaller the letter, the less the drift there is in resistance value as the part gets hot. Advertising "military grade" doesn't really mean anything. Actually, if you read the military specification you'll find that it's just that; a specification. It has to do with the parts physical size, and its minimum performance requirements. Beyond that, it is not an assurance of the part's ultimate quality.

METAL FILM NUMBERING

Because of their accuracy, metal films are available in many more values than other types, and require a more accurate way to label that value. The military numbering system consists of four numbers printed on the body. The first three are value numbers, with the fourth being the number of zeros. 4752 would equal 475 00, or 47,500, or 47K5. Carbon comp resistors are not available in this value, 47K being as accurate as they come in their decades, the next largest value being 51K. In metal films, you could get 47K5,

48K7, 49K9, or 51K1. This is very important in filter design, where a precise value is needed to hit an exact frequency, and tolerance is important, when balancing stereo channels.

METAL OXIDE RESISTORS

Next, we have metal oxide resistors, which are similar to metal film, except that a metal oxide film is used. They are 5% tolerance, 1 Watt and larger parts. They are a good choice for power resistors, but a more common choice is wirewound.

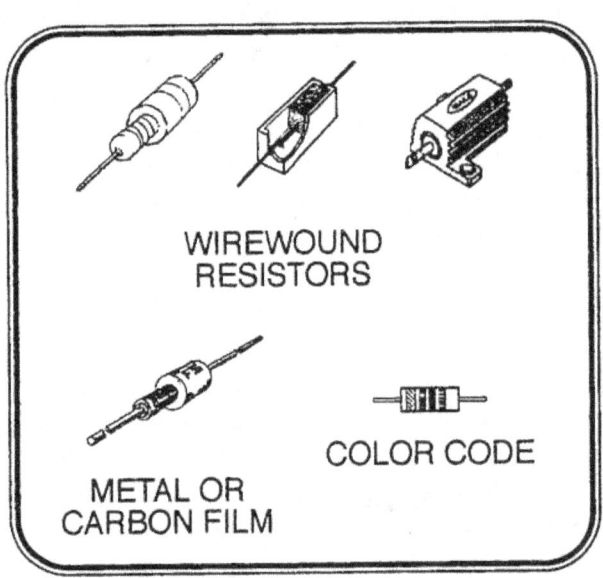

WIREWOUND RESISTORS

METAL OR CARBON FILM

COLOR CODE

WIREWOUND RESISTORS

The 1K, 5 Watt "sandbox" resistors used in Marshalls for screen resistors, are wirewound. A resistance wire is wound around a former and cast in sand (cement) for thermal mass. It is very robust, with its biggest problem being that it is, after all, a coil of wire, also know as an inductor. This is not a problem when used in a power supply, or screen resistor application, where the inductance is actually a benefit. It acts to further filter out

AC Voltage. It would not be smart to use these in a signal path.

BUYING RESISTORS

If you can't find carbon composition resistors, then I would recommend you find an electronic parts distributor, who has the blister pack resistors displayed. These are mostly high quality, 2% carbon film, or wirewound, flameproof resistors. They cost under a dollar for two to five resistors of the same value. If you're wondering why they're called "flameproof," they don't burst into flame when they "burn-up," like carbon composition resistors can.

CHAPTER 6
POTENTIOMETERS

Potentiometers are basically resistors, with an adjustable center point, called a "wiper." They have three terminal connections, numbered 1, 2, and 3. Depending on how these terminals are connected, they can be used as potentiometers, or variable resistors.

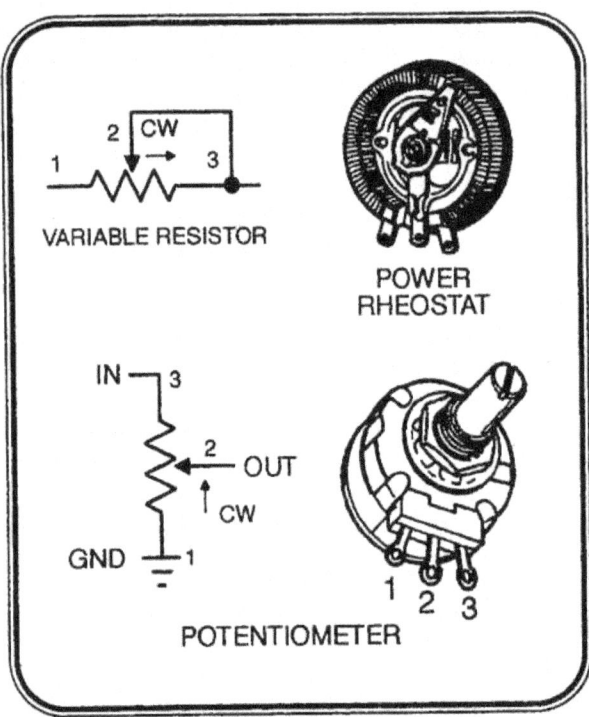

When looking at a pot, you'll find terminals "1, 2, and 3." A signal is applied across terminals "3 and 1." When connected as a potentiometer, terminal "1" is normally connected to ground. The signal goes into terminal "3", and comes out of the wiper, or terminal "2."

Terminal "1" is normally grounded. This "drops" the signal across the pots total resistance, and makes a percentage of it available to the wiper, depending on the wiper's position across the resistance.

When looking at a part drawing in a catalog, always pay attention to the view point. With the normal front viewpoint, it's like looking at the pot from the front panel of your amp. It doesn't matter if the pot is mounted with the terminal "lugs" up, or down, or how the knob is attached. Your ability to "mirror" the part will account for looking at the pot from the back, or from inside the chassis. If this seems stupid to you, you'd be surprised how many designers are right now, sitting in offices, looking at parts, scratching their heads. With most parts being inserted into circuit boards these days, you don't want to be the guy who made the mistake. Don't be afraid to imagine, or even measure the operation of a part to figure out how it should hook up.

When the shaft is rotated totally counterclockwise (knob pointer on zero), the terminal "2" wiper will touch terminal "1," and if terminal "1" is grounded, then the wiper will also be grounded. As you rotate the knob clockwise, the wiper comes off the "peg," and a percentage of the signal applied to terminal "3" will be passed out of terminal "2." The more rotation on the knob, the higher "up" the wiper moves, and the larger the output signal.

VARIABLE RESISTORS

When connected as a Variable Resistor, if terminals "1," and "2" are shorted together the resistance will go down, as the pot's shaft is rotated clockwise. To have more resistance with clockwise rotation, short terminal "2," and "3" together. Sometimes the wiper isn't shorted to either side, and only the wiper and one of the outside terminals is used. This is how most presence, and bass tone controls work. In high Wattage applications, variable resistors are called rheostats.

POTENTIOMETERS

Potentiometers come in sizes, shapes, values, construction techniques, tolerances, Wattages, and "curves," or "tapers." The most common values for pots in guitar amplifiers are:

1,000K (1M)	500K	250K
100K	50K	25K
10K	5K	

The "M" stands for millions of ohms. The "K" stands for thousands of ohms. These letters are sometimes followed by an "A," "B," or "L." These letters stand for the "taper."

POTENTIOMETER TAPERS

There are four "tapers" that you'll see in guitar amps: Linear (U), Analog (A), Logarithmic (A), and Anti-Logarithmic (B). Notice there are two tapers identified as "A." The original "A" is for Analog, and is a 30% taper. Fender amps used 30% pots. Today, logarithmic at 10% is the closest available substitute, 30% being out of production, unless custom made. Many schematics will show a 1MA pot, meaning "analog." If no second letter is shown, assume Linear. If 100KL is shown, it means 100K Linear. I've never seen the part number code "U" used to denote linear pots in a schematic. The Anti-Logarithmic "B" taper is used in vibrato circuits. But what is a taper?

To explain tapers, let's start with the easiest, Linear. Imagine if you connected terminal "1" of a 100KL pot to ground. If you attached an ohm meter between terminal "1" and terminal "3," you'd measure 100K ohms. This doesn't vary as you rotate the pot, so let's move on. Attach the ohm meter between terminal "1" and terminal "2." With the pot rotated full counterclockwise, you'd measure zero ohms. At 50% rotation, you'd measure 50K ohms, and at 100% rotation, you'd measure 100K ohms.

With an Analog taper you'd measure 30K ohms at 50% rotation. This is a 30% taper. With a Logarithmic taper you'd measure 10K ohms at 50% rotation. This is called a 10% taper. Remember, we're measuring the resistance between the wiper and ground, and it's increasing as the pot is rotated up. With an Anti-logarithmic taper you'd measure 90K ohms, at 50% rotation, and it would continue to increase with more rotation. This is called a 10% reverse taper.

GANGED POTS

Pots can be "ganged" together to form multiple sections. Usually, the sections are the same value and taper, and can be controlled by a single shaft, or separately by "concentric" shafts. The outside concentric shaft controls the front section, while a smaller, inside shaft reaches through to control the back section. Sometimes a switch is attached to the back that operates either by rotating the knob off of zero, or by pushing, and pulling the knob in and out.

A dual section pot is usually connected to operate as a master volume control. It would be wired so both sections increase their resistance with clockwise rotation. Of course, I'm talking about the resistance between terminals "1 and 2." Meanwhile, the resistance between terminals "2 and 3" is decreasing. Keeping this in mind, the sections could be hooked up so one increases, while the other decreases (Pan Pot). Just reverse terminals "1 and 3" on one of the sections. Remember, if you're using tapered pots, one of your curves will be backwards.

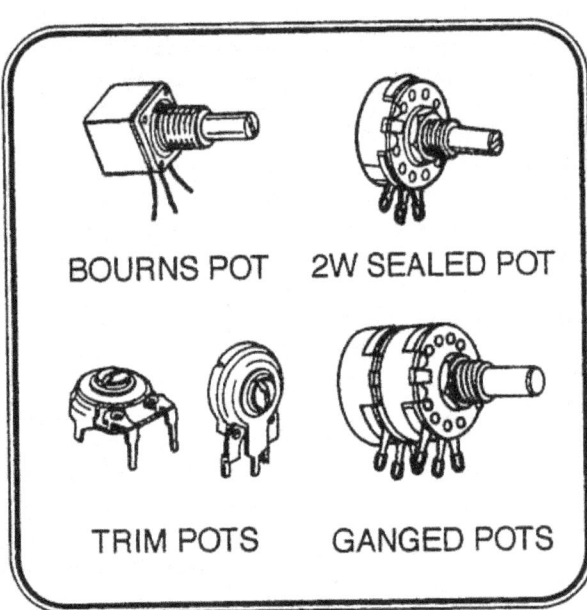

TYPES OF POTS

Besides values, tapers, and Wattage, pots also come in different sizes, shapes, construction types, and tolerances. Tolerances get complicated, but relate to total resistance tolerance, taper tolerance, and tracking tolerance.

Construction types can be conductive plastic, wirewound, or cermet (ceramic/metal). Most 1/2 Watt pots are conductive plastic, with 2 Watt sizes usually being wirewound. Cermets are usually 1 Watt, and have the lowest noise, with the best feel and durability.

As far as you're concerned with replacement parts, the size and shape of a pot has more to do with mechanical restraints (does it fit?) than anything else. The biggest variable you'll have a problem with is bushing length, and shaft length. Since part stores don't really care about any particular market, they carry "modular" pots, with different shafts that snap in. These are normally pretty junky, and will cost you $3 to $5. You can order pots from Mouser but you'll find they're not of very high quality, and they've got a smaller diameter bushing that won't fit the Fender 3/8" holes without floating around. They are less than $1 each.

The "CTS" pots that Fender used weren't sealed, and let dirt get in the back. These are available for $3-5.

I use the Bourns pots, and have the popular values in stock. They can be adapted to hand wired or circuit board amps, or in guitars. I also have some 250KL and 500KL Allen Bradley Ganged pots that are perfect for a Master Volume control, and most values of the Allen Bradley or equivalent single pots. They're not cheap, but after the Bourns, they're the best.

CHAPTER 7
COILS, CHOKES & TRANSFORMERS

A coil of wire is the electrical opposite of a capacitor. Capacitors like to pass high frequencies, and they resist low frequencies. Coils like to pass low frequencies, and they resist high frequencies.

Whenever electricity flows through a wire, a magnetic flux field is created around the wire. This field changes when the flow of electricity changes. If the flow of electricity reverses itself, the field collapses, and reforms itself in the opposite direction. This building, collapsing, and rebuilding of the magnetic field, takes a finite amount of time. As the frequency increases, the field can no longer reverse itself fast enough. This causes an increase in the coil's resistance to electricity at these higher frequencies. Because this resistance varies with frequency, it's called an impedance. Because this impedance is from a coil of wire, and by definition it increases with frequency, it's called inductance. The longer the coil of wire, the greater the inductance. Even a straight wire has a small amount of inductance. Inductance is measured in Henrys. Smaller values are measured in milli-Henries (mH), or micro-Henries (uH).

1.0 H = One Henry
.001 H = One milli-Henry or 1 mH
.000,001 H = One micro-Henry or 1 uH

Coils of wire also have a DC Resistance (DCR). This is the coil's resistance to Direct Current. Since Direct Current has a frequency of zero, the coil's inductance

doesn't affect it, but its DC Resistance does. For a constant inductance, a coil's DCR can be made smaller by winding it with a larger gauge of wire. This reduces the IR drop (Voltage lossed) through the coil, while retaining its impedance to frequency fluctuations. The larger wire size would, of course, increase the size of the coil.

The most common use of coils in vacuum tube amplifiers is the choke in the power supply. Its the smallest of the three "transformers" you'll see on the chassis. The one with only two wires coming out of it.

Chokes are coils of wire used to filter out (choke, restrict) the 60, or 120 cycle fluctuations in a power supply. In vacuum tube amps, a choke is normally connected in the high Voltage DC supply, between the power tube Plates and the screens.

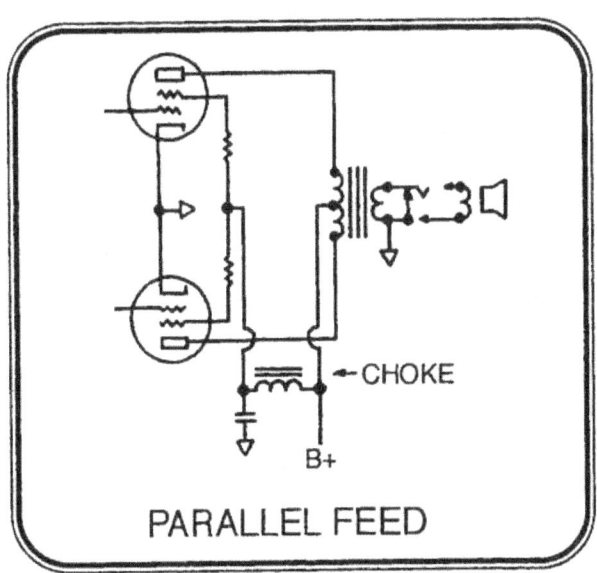

PARALLEL FEED

49

Whereas filter caps are connected from the B+ supply to ground (parallel, in shunt, across the supply), chokes are connected in line (series) with the B+ supply. Remember, coils are the opposite of caps, which should explain their opposite connection in the circuit. We'll get more into power supply design in a later article.

The chokes used in power supplies are quite large to efficiently pass the large DC Currents, while filtering the low frequency AC component to Ground. The schematic symbol for coils is "L." The combination of coils and caps is known as an LC filter. If the coil comes first, it's known as a choke-input filter. If the cap comes first, it's known as a capacitor-input filter. The choke in a Fender amp is rated at 3H, @ 50ma, with 92 ohms DCR. The English like to use even larger chokes, like the 15H choke in old Vox amps.

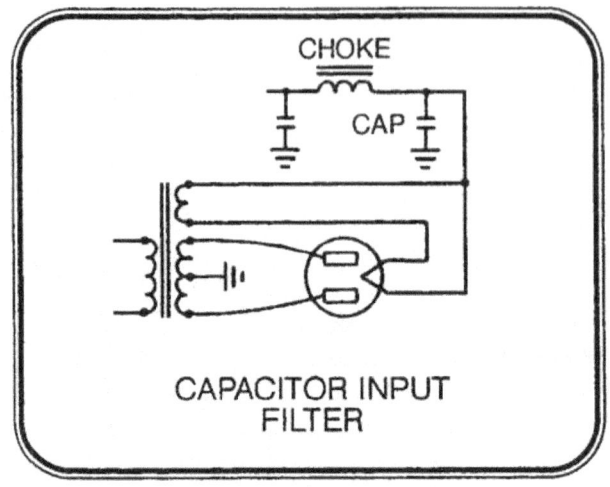

CAPACITOR INPUT FILTER

CAPACITOR INPUT FILTERS

Some of the earlier Fender amps had Choke Input power supply filters, but the most common is the Capacitor Input Filter. After the rectifier, there are two 70/350v caps in series (look at the chapter on Vacuum Tube Diodes). The center tap of the output transformer is connected to these caps. The Plates get the effect of a choke filter because the B+ must flow from the center tap of the output transformer through the coils of the output transformer to the Plates.

Any hum induced is also cancelled in a push-pull design by the opposition of the two sides. The B+ Voltage also flows through the choke to the second capacitor filter stage where the power tube screens are attached. The screens have the benefit of the first set of caps, the choke, and

the second set of caps.

Choke filters are efficient, but the chokes themselves are expensive. In smaller amps, a resistor is sometimes used in place of the choke. The resistor provides poorer Voltage regulation, which, due to the voltage sag on the Screens, results in compression. I'm specifically talking about vacuum tube amps where the choke/resistor segment of the filter is located between the Plate supply and the screen supply. To understand this, I'll need to explain something about pentodes.

The B+ Voltage on the Plate sets the potential power you can get out of the tubes, but the screen Voltage is the tube's "throttle." As the signal through the tube gets louder, the screens draw more, and more Current. This Current comes through the choke. The choke's DC resistance causes an IR drop (a drop in Voltage equal to the Current in Amps, multiplied by the choke's resistance in Ohms). The IR Drop lowers the Voltage on the screens. The lower screen Voltage reduces the gain of the tubes. It's like lifting your foot off the throttle everytime a big signal comes through the amp. This is the very definition of compression. Some amps with solid state rectifiers use this trick to simulate the Voltage sag of a tube rectifier, by replacing the Choke with a large resistor. About 1K to 10K ohms at 5 to 25 Watts is a good starting point.

The screen resistors have the same effect. You might try checking the effect of the 2K7 screen resistors on newer Marshall amps.

CHOKE INPUT FILTERS

Some early Fender tweed amps have choke-input filters. They generally provide better Voltage regulation, and protect the rectifier tube by restricting the inrush of Current when first turned on, but they have lower output Voltage.

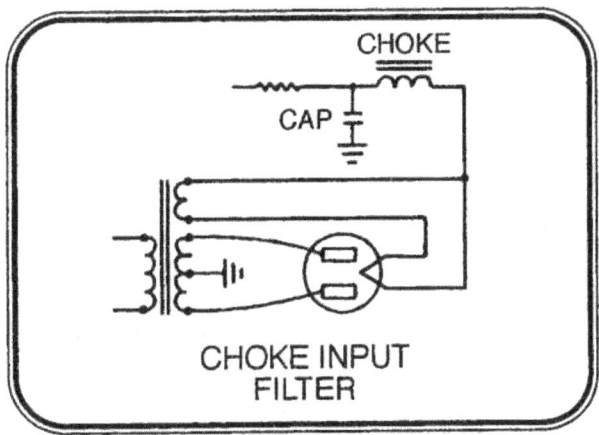

CHOKE INPUT FILTER

After the high Voltage AC from the power transformer secondary goes through the diode rectifier, it comes out as high Voltage DC with a large amount of DC ripple on it. If this is passed through the choke, there is a large loss of Voltage as the choke resists the large DC ripple. Remember, the DC resistance of the choke causes an IR drop in the DC Voltage, but in addition, the ripple compo-

nent of the Voltage is ALSO resisted by the choke's impedance (measured in Henries). Where there might be 90 ohms of DC Resistance, there could be 2200 Ohms of Inductive Reactance.

A better configuration is to attach a filter cap to ground first. This smooths, and reduces the size of the ripple, and allows the Choke to deal mostly with straight DC Voltage. Now you have much less Voltage lost in the Choke.

TRANSFORMERS

Transformers are made from two coils of wire. They convert (transform) Impedances. For now, let's just say that they convert Voltages. They do this by inductive coupling. If you place two (or more) coils of wire next to each other, and put a signal into one coil, a copy of the signal will be "induced" into the other coil(s). The first coil is called the primary, and the second coil is called the secondary. In some cases primary and secondary are relative terms, meaning the primary and secondary are interchangeable.

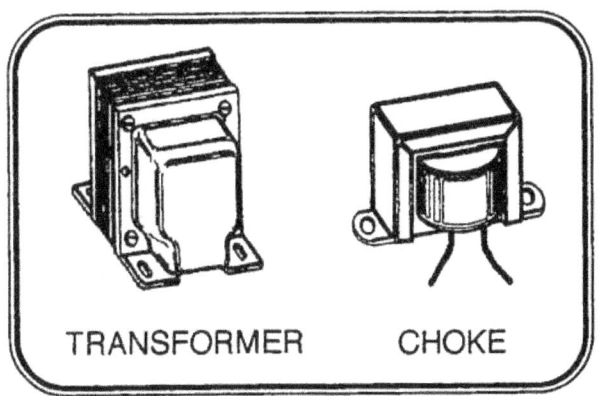

TRANSFORMER   CHOKE

When a Current flows through the primary, a Magnetic Flux field is created, and this field transfers a Voltage into the secondary, but this only happens when the field is changing. If you attach a 9 Volt battery to the primary of a transformer, a pulse will be induced into the secondary, but only when the connection is made. When a steady 9 Volts of DC flows through the primary, no signal, after the initial pulse, will transfer. From this you should see that transformers, like capacitors, pass AC, but not DC (except for DC "pulses").

If both coils of wire have the same number of turns, the Voltage induced into the second coil of wire will equal the Voltage in the first coil. This would be a turns ratio of one to one. Turns ratio is how transformers "transform." By varying the number of turns in the primary, or secondary, different output Voltages can be created. This is how power supply transformers create the higher, and lower Voltages needed by an amp.

If the primary coil of a transformer has 100 turns, and the secondary has 200 turns, the output Voltage will be twice that of the input Voltage. If the primary is connected to 120 Volts AC, the secondary will have 240 Volts of AC available on it. This works regardless of the actual number of turns in the windings. What's important is the ratio of turns between the primary, and secondary.

OUTPUT TRANSFORMERS

An Output Transformer matches the tube's Plate Impedance to the Speaker's Impedance. In doing so, it converts the tube's High Voltage and Low Current, into relatively Low Voltage and High Current. Many tube amps have impedance selector switches. These switches select a differ-

ent point, or tap, on the secondary winding to connect to the speakers. These different taps change the turns ratio, and therefore the relative output impedance.

If you remove two tubes from a 100 watt amp, the Output Transformer's Primary Impedance needs to be doubled. To do this, double the Impedance of the speaker load. In other words, four tubes might need a 2000 ohm Primary, with a 4 ohm speaker load. By removing two tubes, the OT's Primary Impedance now needs to be 4000 ohms. By attaching an 8 ohm speaker, you've effectively doubled the Primary Impedance. Remember, it's a ratio. An example of this was the Triad transformer #45548, used in the 6G8 Twin Amp. It's specification calls for 4,000 ohms of Primary with an 8 ohm speaker load. Since the 2-12" actually had 4 ohms of Impedance, the Primary became 2,000 ohms. This matched the loads, but the transformer was only rated at 40 watts. The other transformer used on the 5F8 Twin and the 6G8-A Twin was #45268, rated at 2,000 ohms to 4 ohms at 80 watts. These Transformers are easy to tell apart, since the 80 watt one has twice the iron in it. I wonder if this isn't just a typo on the 6G8 schematic. Let me know if you find a #45548 in your 6G8 Twin.

A common source of transformer noise is vibration caused by the Current flow in the coil windings. Many transformers are vacuum filled with wax to prevent this. The transformers are placed into a large vat filled with wax. Some of the wax soaks in, but to insure impregnation, the entire vat is sealed, and the air pumped out of it. This draws the air out of the transformer's windings, and they become filled with wax.

Vox transformers were filled with beeswax. This is a remarkable substance, but it absorbs water. These transformers are rusting from the inside out.

TRANSFORMER NOISE

Electrical noise is sprayed off the output transformer from the gaps in the laminations. This noise can be reduced by orientating the output transformer correctly. There is a very expensive amp out there that's done wrong.

There are many factors that effect output transformer performance, and this could be a book in itself, but I'll cover some of the basics here.

First, the size of the transformer's core relates mostly to its low frequency response, and amount of distortion allowed. Once the magnetic flux "saturates" the transformer's magnetic core, it will not pass additional signal. The strongest magnetic fields are at the lowest frequencies. Don't be fooled into thinking that large transformers automatically mean they have a higher power rating. They might just have a lower frequency response at a certain power level.

Since a larger core doesn't reach magnetic saturate until a higher power level is reached, which occurs at the lowest frequencies, and saturation causes distortion, you might say a larger transformer can:

      1. Handle Higher Power
      2. At a Lower Frequency
      3. At Lower Distortion

But, the power rating of a transformer has to do with its wire size, and its temperature capability. The larger core, BY ITSELF, doesn't mean that the transformer can handle higher power. Of course the larger wire sizes make for a larger coil of wire, which needs a larger lamination to hold the coil, and it needs a larger magnetic core, making for a larger transformer. So, don't be fooled by what I said in the previous paragraph.

LEAKAGE INDUCTANCE & INTERLEAVING

The high frequency response of a transformer depends on its Leakage Inductance, which varies with the transformer's design. Moving the coils closer together by using a thinner insulator, like plastic rather than paper, or Interleaving the windings, reduces the Leakage Inductance. The lower the Leakage Inductance, the higher the frequency response, up to a point.

Another common way to reduce Leakage inductance is to "Interleave" the windings. This means to wind part of the Primary; then stop and wind some of the Secondary. Continue on with more Primary, and then more Secondary. The number of times the coils are woven together is called the number of Interleaves. The effect is to move the coils closer together. For all the hype about Interleaving, it may surprise you to learn that the Holy Grail of Output Transformers, the 4-10 Bassman's, had — are you ready? — only 2 Interleavings. Thunderfunk output transformers are presently not interleaved at all. I get 12KHz-14KHz of response out of them without interleaving. Since the speakers only go to 6KHz, what's the problem? It has to do with having a good transformer company build you a good transformer. When you hear hype, it's because the person doing the hype really has nothing else to offer to justify outrageous prices.

Another thing to watch for is the use of paper to insulate transformers. Just because they did it in the old days, and it might be a "Secret," doesn't make it a good idea. The Triad transformers used by Fender (they're really Fender custom transformers built by Triad, and were not available to the general public), were only for breakdown tested to 1,500 Volts. Thunderfunk transformers are tested to 4,000 Volts.

There's further B.S. out there that paper is a better insulator than plastic, and therefore can be thinner. Paper is a better insulator than plastic? Again, the Fender specification calls for .021" of Kraft wrapper between layers. Thunderfunk transformers use .007" of mylar. Maybe because the coils are .014" closer together, the transformers don't have to be Interleaved.

SPEAKER MATCHING

For the maximum power out of a tube amp, you need to attach the right speaker load. If you can't match exactly, it's better to have a higher speaker impedance, rather than a lower one. Expect a 12% power loss with a mismatch of one step, and a 25% power loss with two steps of mismatch (a 16 ohm cabinet on a 4 ohm transformer). This, by the way, sounds terrible. If you want to see for yourself, put a 16 ohm 4 x 12" cabinet on a 4 ohm Bassman head. And, it's hard on the amp.

By the way, again, there's a joke going around about series-parallel, and parallel-series wired Marshall cabinets sounding different due to "distributed inductance." Someone should go back to school, and maybe study electronics, or even learn math, instead of wasting his time playing guitar in a bar.

TRANSISTOR AMPS

Transistor amps generally don't have output transformers, or impedance selector switches. You'll also notice that transistor amps put out more power as the impedance of the attached speakers goes down. The lower speaker impedances more closely match the transistor's low internal impedance, resulting in a better transfer of power. The best power transfer occurs when equal impedances are connected. Since the transistor's internal impedance is not adjustable, the lower speaker impedances result in a better match; a better transfer of power; and higher Wattages.

While we're talking about matching speaker loads, let's talk about power soaks. The problem with most power soaks, is they mismatch the load impedance that's taking the place of the speaker. This is how they can ruin your amp. It is possible to build one that dissipates the excess power, and correctly matches the impedances. These are usually called "L" pads. I'm confident, although I haven't checked, that the power soaks made by Trainwreck, and THD are excellent products.

Now, on to the tubes themselves.

CHAPTER 8
VACUUM TUBES

Now that we've covered what are known as the "passive" components - resistors, capacitors, and inductors - we're ready to get into amplification.

Amplification changed electrical engineering into electronic engineering. Before amplification, you could only subtract from a signal with passive components. With amplification, you could invent devices that need gain; like the radio, radar, tape recorders; even guitar amps.

Amplification is the definition of an active circuit. Active circuits have gain. Gain is the ability to make a small signal bigger. The first device to do this was the vacuum tube. In future articles, we'll also cover transistors and operational amplifiers, and we'll discuss amplifier circuits, and modifications you can make.

THE UNIODE

The first vacuum tube was the light bulb. I guess you could call it a uniode. It has one element inside of it, called the filament. If you hook it up to a current source of the proper voltage, it will get hot, and glow with light.

All tubes have a filament (except "cold cathode" tubes). The first material used was made from pure tungsten. In order to attain the dazzling white high temperatures the tungsten needs to emit sufficient electrons, a relatively large amount of filament power is required. To reduce the power demands, the tungsten is impreg-

nated with thorium oxide. These filaments emit electrons at a more moderate temperature of about 1700C (a bright yellow) and therefore, are much more efficient in their use of filament power.

To further reduce power demands, alkaline earths can be applied as a coating on a nickel-alloy wire or ribbon filament. This coating, which is dried in a relatively thick layer, requires only a relatively low temperature of 700-750C (a dull red) to produce an abundant supply of electrons. Coated filaments are very efficient, and require relatively little filament power.

THE CATHODE

A Cathode is a Negatively charged element in a vacuum tube. The Filament itself, sometimes serves as the Cathode. In a one element tube, the word Cathode is irrelevant, because there is no complementary positive element in the tube, which would be called the Anode. So the word Uniode is really a misnomer, since without two elements, it really can't be an "-ode" at all.

In some tubes the filament, and the Cathode are separate pieces. This is called an Indirectly Heated Cathode. In these tubes, the filament is only called "the heater."

Directly heated Filament-Cathodes require comparatively little heating power. They're used in tube types designed for battery operation. An example of a Directly

Heated tube is the 5Y3 GT rectifier.

When you plug you amp into the wall it called, "AC operated." With AC Operation, there's no need for the efficiency of a directly heated cathode. Enter the Indirectly Heated Cathode, consisting of a thin metal sleeve coated with electron-emitting material such as alkaline-earth oxides, with an insulated Filament wire inserted to heat the Cathode Tube from the inside.

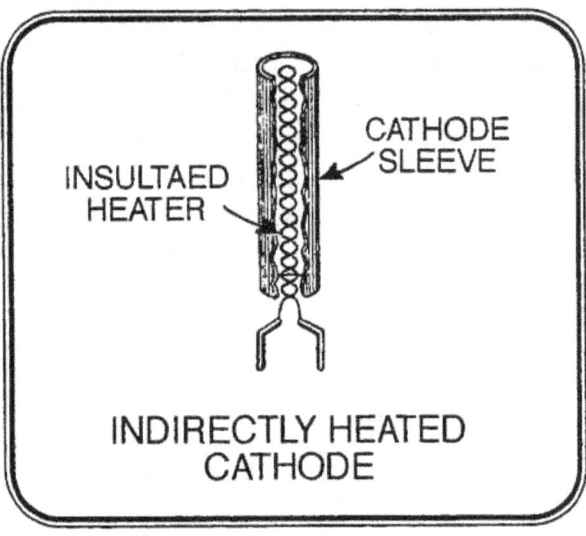

The use of separate pieces for the Cathode and Heater; the fact that they're electrically insulated from each other; and the shielding effect of the Cathode Tube over the Filament, all may be utilized in the design of the tube to minimize the introduction of hum and electrical interference from the AC heater supply.

Another advantage of the Indirectly Heated Cathode is that it allows close spacing of its Cathode and its Plate, reducing the tube's Impedance. In a Power Supply, Indirectly Heated Cathode Rectifiers provide better Voltage Regulation (less sag), due to their lower Impedance, and lower Voltage drop

through them. (Same thing).

In Amplifier Tubes (3 elements and more), the close spacing increases the gain obtainable from the tube. An example of this is the 6550, which has close Grid to Cathode spacing, but therefore, is also prone to Grid to Cathode shorts.

Almost all the tubes you'll see in guitar amps, were (are) designed for AC operation, and have Indirectly Heated Cathodes,

FILAMENT VOLTAGES

The first number of a tube type's number refers to the needed filament voltage. A 6L6 tube is designed for 6 volts (actually 6.3 volts). A 12AX7 needs 12.6 volts. In the old days, you might find a 50C5, a 35W4, and two 12AU7's in the same amp. If you hooked up the 50 volt tube in series with the 35 volt tube, and the two 12 volt tubes, it adds up to 109 volts. Now you can connect the filaments directly to the 110 volt wall outlet, without needing a filament transformer. It was one way to save money. The only problem is if one filament burns open, all the tubes go dead. Then it's hard to determine which tube is bad without some type of test equipment, or having a spare of each different tube type.

Many 12 volt, nine pin minature tubes have two 6 volt filaments connected in series internally. That's why in Fender amps the 6.3 volt green filament wires connect to pin 9, and pin 4 & 5 are soldered together for the second wire. Pin 9 is the middle of the 12.6 volt filament. By joining them, and using pin 9, the filament is divided in half, and the required filament

56

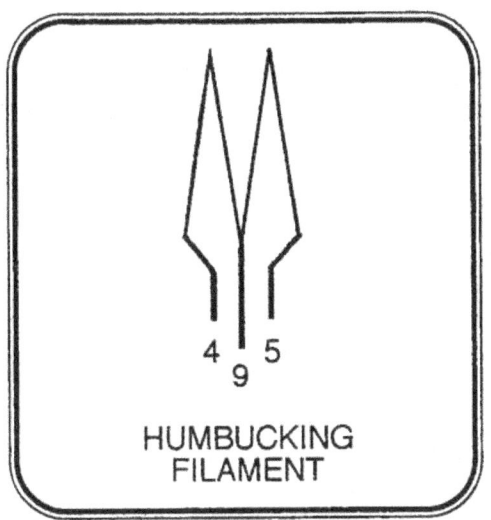

HUMBUCKING FILAMENT

voltage is 6.3 volts. If you separated pins 4 & 5, and attached the two green wires to them, you'd need 12.6 volts. The tube was designed to run on 6.3 volts and the twin filaments are there to reduce AC Hum pickup from the Heater's supply.

Different tubes require different amounts of filament current. I get calls from guys wanting to change 6V6's to 6L6's, and 6L6's to 6550's. You need to check the heater current requirements before changing. A 6V6 needs 450ma. The 6L6 needs 900ma, and the 6550 needs 1800ma. Some 6V6's are actually 6L6's. They should not be interchanged. The Princeton and Deluxe Power Transformers are rated at 2 Amps of 6.3 Volt Filament Current. If you change to a higher tube, you could overheat the filament winding of the power supply transformer, and burn it out. If you do change, it's best to add another, small 6.3 volt transformer to heat the power tubes, or preamp tubes separately. The 12AX7 type tubes need 300ma each at 6.3 volts, or 150ma at 12.6 volts. If you're going to do this, it might be cool to rectify the output of the new transformer to make DC, and use it to light the preamp tubes.

THE DIODE

Early light bulbs had a problem with the inside of the glass turning black as the filament burned. The diode was accidentally invented when a Plate, called the Anode, was inserted into the bulb as an experimental way of keeping the bulb from turning black. Not knowing what they were doing, a positive Voltage was applied to the "Plate" to increase its attraction of the black substance, and it was noticed that a Current flow was created between the Filament and the Plate. This was a real surprise, as no one expected that electricity could flow through a vacuum. This was the invention of the "diode."

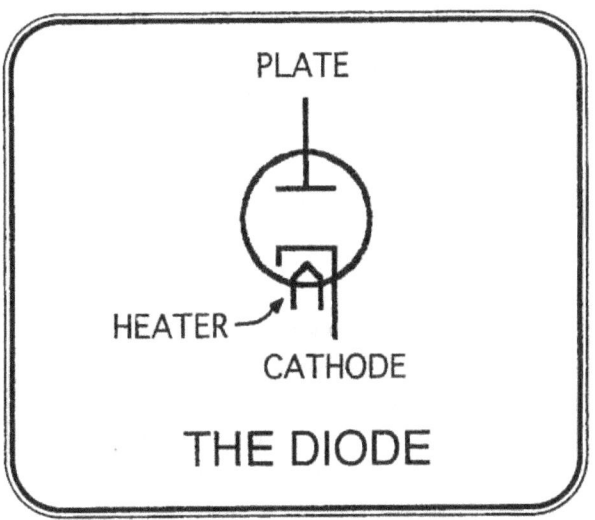

THE DIODE

The way it works is the Filament (or Cathode) emits electrons that are attracted by the "Plate" (the Anode). The amount of Current flow established is based on the temperature, material, and physical construction of the Cathode; the amount of positive Voltage on the Plate; and their physical distance from each other. These factors create a certain amount of Current flow that stabilizes based on these factors. In other words, a two element tube stabilizes to some level of a full on condition. Beyond a certain point, no increase in Plate Voltage will increase the amount of Current flow through the tube. This is called the Saturation Current, or Emission Current. Operating the tube in this area will destroy it, or at least permanently degrade its performance. If you have a tube that's ever gone "cherry," you should replace it.

Also, diodes only conduct Current flow in one direction. You can think of diodes as uni-directional wires. Electricity can flow forward through the wire, but not backwards. Remember, the Current flows from the Cathode to the Plate, but not backwards. Now we need to review the differences between AC and DC electricity.

AC/DC (NOT THE BAND)

Direct Current flows in one direction. Alternating Current flows forward, and alternately, backwards. Diodes can be used to convert AC to DC by blocking the reverse flow of Current in what is called a rectifier circuit. A diode would pass all of the "forward" flowing Current, but would block any "reverse" flow of the AC. Forward Current flow occurs when the "anode" is more positive than the "Cathode." This condition is called "forward bias." No Current flows when the Cathode is more positive than the anode. This is called reverse bias.

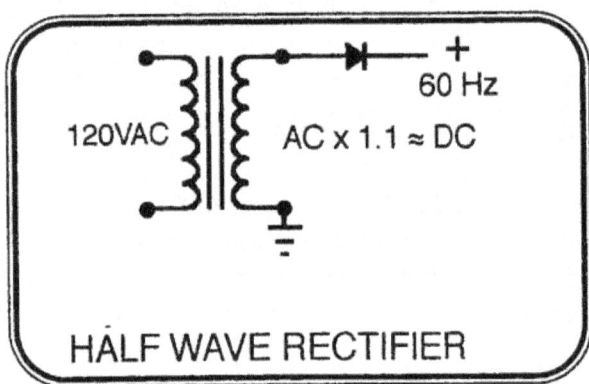

HALF WAVE RECTIFIER

In the United States, the AC in the wall pulses at 60 cycles per second (cps is now called Hertz, or Hz). If you attach a transformer to the AC wall Voltage, the Voltages in it also pulse at 60Hz. If you connect one wire of a transformer's secondary to ground, and the other wire to a diode's anode (Plate), Current will flow whenever a circuit is completed by connecting the Cathode to ground (usually through a "load" impedance), and the Plate is positive with respect to the Cathode.

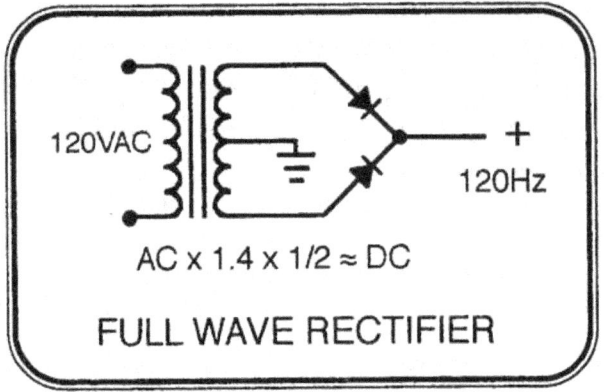

FULL WAVE RECTIFIER

In a Full Wave Rectifier circuit, both the top and bottom of the AC Voltage appear above ground. This is because the top diode conducts on the first half of the 60Hz AC Current, and the bottom diode conducts on the second. The pulses are added together where the diode's Cathodes meet. The result is DC with a lot of ripple on it. This type of rectifier puts out 120 cycle hum. The center tap is needed for two reasons. First, without it, the transformer would have no ground reference, and would gyrate around itself, developing no Voltage relative to ground. Second, it forms the center of the 60 cycle wave. The first half of the AC pulse goes positive from the center tap to the top diode; and the second half of the wave goes positive from the center tap to the bottom diode.

58

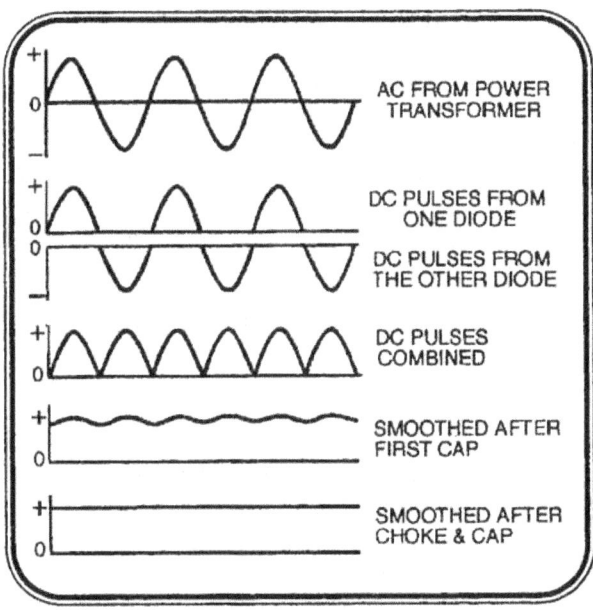

	AC FROM POWER TRANSFORMER
	DC PULSES FROM ONE DIODE
	DC PULSES FROM THE OTHER DIODE
	DC PULSES COMBINED
	SMOOTHED AFTER FIRST CAP
	SMOOTHED AFTER CHOKE & CAP

ATOMIC THEORY?

For those of you who are understanding these chapters, a lot of things are implied. If you grasp what I'm talking about, you can take it to another level of understanding, and learn some amp "secrets" along the way. This section is to explain some of the "relativity" of electricity. There would be no positive without a negative, even if it's only "relatively negative." Sometimes things are described backwards to make things easier to understand. Anyway, if you're already having a problem understanding this book, you might as well skip this part.

Current flows from the ground UP. People tend to think of Current flowing from the "positive" side of a battery, through a load to ground, where it's dissipated. In fact, Current flows out of the ground, through the load, to the positive side of the battery. Also, electricity is not dissipated, or destroyed, but rather, it's Voltage differential is destroyed, and turned BACK into heat, and/or kinetic energy (the motion of a speaker cone). The kinetic energy of

falling water, or the heat of a steam generator, is transferred along wires into your house to push air into your ears? Some concept, huh? Electricity is just a convenient way of transferring power.

Einstein showed that mass is energy in motion. The theory of relativity asks, "How big is an electron?" And, the answer is, "When?" How do you measure something in motion? When do you start measuring it, and when do you stop? The shorter the measuring time, the smaller the electron becomes, until if you could freeze time, all you'd find is energy. Without the motion (also known as time), there's no mass. $E=MC2$. The speed of light in the equation is a time base.

The best definition I ever heard of time was:

"Time is what happens,
when nothing else does."

The detector Plates I helped design for the Proton/Anti-Proton collision experiment at Fermilab were little trapezoids of copper spread out on a 5-foot pie shaped printed circuit board. Each patch of copper had a conductor running to a data connector at the center. These boards encircled the path of the collision. As the protons collided, debris would fly through these patches, and by calculating the path, electrical potential, and speed of these particles, their mass could be calculated, and a determination was made of what they were. In order to capture these very small and fast particles, a super-computer was needed. The physicists at Fermi built their own. Ultimately, the question of time resolution limits our knowledge. Did you ever think that in order to make a 100MHz Pentium computer, you need to test the processor? The test equipment has to run faster than the device under test, and then you have to test the test equipment. If you think the products are cool, you should see the labs.

Energy is used to create electricity by stripping electrons off a conductor. The more electrons stripped off, the higher the Voltage. These electrons leave positive holes behind. If you have a lot of holes, and only a few electrons, you have a hell of an attraction, and a high Voltage. One thing that creates a shortage of electrons, is a high resistance to the flow of those hole canceling electrons. Voltage is developed across a resistance. Without resistance, you have a short; the electrons recombine with the holes; Current flows; and the Voltage disappears.
Now, by inference you should see that a high impedance circuit develops, and is based on Voltage; and low impedance circuits are based on Current flow. Vacuum tubes are high impedance devices (Voltage based), and transistors are low impedance devices (Current flow based).

You should also see that, if the impedance is lower than the circuits ability to supply holes, the Voltage collapses as the positive holes become filled with the readily available electrons, and the circuit comes into balance at a new lower Voltage.

If you have a device supplying a Voltage signal into the next stage, and the driving device can't supply enough Current because of its high internal impedance, and the driven stage has a low input impedance, the driving device can't maintain the integrity of the Voltage wave. The signal peaks will collapse to a Voltage that balances with the next stage's input

impedance, and the available Current.

Do you know what a buffered effects loop is? It supplies enough positive holes to maintain a signal Voltage's integrity into a low input impedance. Of course, that means a "relatively" low impedance. And, of course, a low impedance is a high load, and a high impedance is a low load, relatively.

Now, do the electrons flow to the holes, or do the holes flow to the electrons? It's like the question of whether "light" is a wave, or a particle. Experiments prove it's a wave. Experiments also prove it's a particle. Actually, Current is the flow of electrons to fill positive holes. But, sometimes it's easier to grasp a concept by looking at it "backwards."

I hope this little diversion serves to explain why my chapters sometimes don't make sense to you. It's difficult to explain things in black and white, when sometimes you're looking at a photograph, and sometimes you're looking at the negative used to make the photograph.

RECTIFICATION

On the next page you'll find a chart of the Current and Voltage flows for both positive and negative Voltages. This is where relativity comes in. Remember, a Cathode emits electrons that are attracted by the anode (vacuum tube Plate). When electrons are attracted away, positive holes remain, creating a positive Voltage. When the electrons flow forward, they create a surplus of electrons, resulting in a negative Voltage charge. The negative Voltage drawing shows a NEGATIVE Voltage flowing out of the diode. Notice, in both

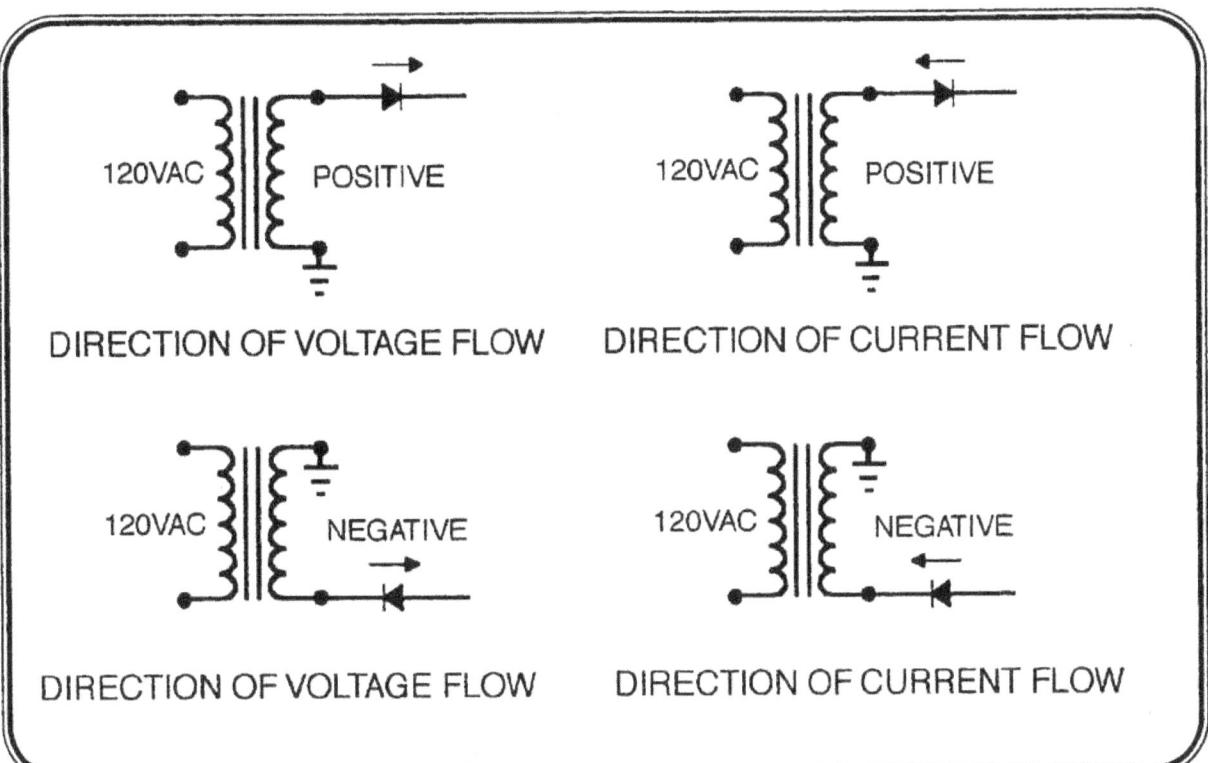

DIRECTION OF VOLTAGE FLOW   DIRECTION OF CURRENT FLOW

DIRECTION OF VOLTAGE FLOW   DIRECTION OF CURRENT FLOW

cases, the negative ELECTRON flow is shown from Cathode, to anode. That is the only direction that Current flows through a diode. The uni-directional nature of diodes is what makes them special. The argument of whether the negative electrons flow or the positive holes flow, is one you'll hear. To keep things simple, we'll always consider the negatively charged electrons flowing.

When a solid state diode is drawn in a schematic, the Cathode is indicated by a flat line. On the physical parts, a white stripe indicates the Cathode. On a round LED, the flat spot is the Cathode, and it has a longer wire lead. NOTE: If something doesn't work right, don't assume the part is marked correctly.

Remember, in order for electrons to flow, the anode has to be more positive than the Cathode. For positive Voltage, this

happens during the AC cycle when the transformer winding connected to the diode's anode is more positive than the Voltage on the other side of the diode, beyond the Cathode.

In the negative Voltage drawing, the Current flows, again, when the anode is more positive than the Cathode. Since the Cathode is connected to the transformer winding, the anode is more positive when the transformer winding is on the negative leg of the AC cycle. Current, then, will only flow during the negative cycle. That's why I've drawn it "upside down." To illustrate the negative leg of the AC cycle, which is below ground. Now, in practical terms, which end of the winding is grounded, and which end attaches to the diode is irrelevant, EXCEPT in transformers that have a connection between the primary and the secondary windings on the ground side. If there's no connection, the transformer is isolated, and is called an isolation transformer.

I once had to connect two transformers together to produce enough Current for an experiment. In that case, I used an oscilloscope to determine the correct phase of the two transformer windings I was connecting together. This allowed the transformers to both "push" at the same time, doubling the Current available.

The astute will see that in a power supply, the largest Current flow occurs when the filter capacitor attached to the diode's Cathode is empty. As the cap fills with positive Voltage, the duration of the AC cycle when the Anode is more positive than the Cathode, is reduced. With vacuum tube rectifiers, there's a limit to how large of a filter cap can be connected to it. Larger caps have a lower impedance. When the amp is first turned on, the large cap can cause the rectifier tube to conduct excessive Current, and burn out. One way to control this, is to use a "Choke Input" Filter. This circuit places a Choke (coil) before the first Capacitor. The turn-on surge is damped by the Choke.

NOTE: This is also an easy way to lower the DC Voltage in an amp. You'll need a larger Current capacity Choke than the one used between the Plates and the Screens. The one that comes with the amp doesn't have to pass the Current used by the power tube's Plates. I'd suggest a minimum size of 1 Henry at 125 to 300 milliAmps.

TUBE RECTIFIERS

Many old tube schematics are drawn "roughly." No pin numbers, and not always showing every component. This is where your experience comes in. You're not going to really know how a tube amplifier works, until you can draw an entire schematic from memory, and explain what every part does. If pin numbers are left off the schematic, you'll have to refer to a tube manual, or your memory.

The three most common rectifier tubes are the 5Y3, the 5U4, and the 5AR4, also known as the GZ34. They all require two things; an AC Voltage to be rectified into DC; and 5 Volts AC to light their heaters. These tubes have different Current requirements for those heaters. They also have different maximum Voltage and Current ratings for their AC Plates, as well as their DC output. A tube might have a higher rating in one area, and smaller in another. So, which tube to use depends on your application.

Also, the data book doesn't have a "standard" set of specifications for tubes. When you try to compare one to another, you'll find some ratings omitted, or specified differently. The one thing you can say, is the 5U4 is the biggest of the three, both physically, and electrically.

| | HEATER | | AC PLATE | DC OUTPUT | |
	VOLTS	AMPS	AMPS	VOLTS	CURRENT
5Y3	5v	2.0	400ma	500V	125mA
GZ34	5v	1.9	750mA	550V	250mA
5U4	5v	3.0	1000mA	550V	275mA

REPLACEMENT TUBES

You should only use the correct replacement tube in an amplifier. If you try switching things around, you'd better know what you're doing, or be prepared for damage.

Some amps take the high Voltage DC off pin 2 of the rectifier (Gibson GA-8). This is fine with a directly heated rectifier, but a GZ34 is indirectly heated, and pin 8 is the Cathode connection. If you plug a GZ34 into the amp designed for a 5Y3, with the DC taken off pin 2, you'll be pulling the DC Current off of the heater, instead of the Cathode. It should work, for awhile. If you work on an amp like this, you should probably move the DC wire to pin 8, to avoid someone making this mistake in the future.

Also, the original Vox AC30 was designed for a GZ34, and that's the correct replacement tube. Most other rectifiers will not physically fit the chassis! The GZ34, being indirectly heated, will also warm up slowly, negating the need for a standby switch. Directly heated rectifiers will pass DC within a couple of seconds, before the power tubes are warm enough to accept it, causing the Cathodes to be stripped of their oxide coating, and bringing about premature tube failure.

RECTIFIER TONE

Why do tube rectifiers sound different? It's because they have relatively high, frequency sensitive, internal impedances. These impedances cause Voltage loss during power demand (sag), and because the "sag" varies with frequency, a "resonance" is created. In other words, amplifier circuits are alive. Every part interacts with every other part. If you ask, "How much sag will I have?" The answer is: 1. At what power level?; 2. For what duration?; 3. At what frequency?; 4. With what

rectifier?; 5. With what power tubes?; 6. With how much filter capacity?; 7. What's the size and DCR of the choke (if there is one); 8. What's the regulation and DCR of the transformer?; 9. At what wall Voltage?; 10. Ecetera.

The point is, the lower the impedance of the power supply, the better. That means solid state diodes, and huge filter caps. So, why do we use tube rectifiers in guitar amps? Here it is. Because guitar amps generate sound. Hi-fi amps reproduce it. You don't want coloration in your stereo system, but you do want added body in your guitar amp.

What do rectifiers sound like? Solid state rectifiers go, "Thonk." Tube rectifiers go, "Thump." Now, don't assume the tube rectifier sounds better. It depends on your touch, strings, pickups, cord, amp, power tubes, speakers, what you're used to hearing, and your style of playing. The different types of tube rectifiers vary in their sonic clarity. The Sovtek GZ34 works fine in Fender tweeds, which have a fuzzier tone, but they don't work as well in old Voxes, which have a more bell like tone.

A step up from the Vox is the Trainwreck Rocket amp, which is a Vox taken to heaven. If you want to compare the sound of different rectifiers, use a Rocket. The amp has such clarity that Ken won't build one, unless he can get a Mullard GZ34 to ship with it. Anything less, and the amp won't perform to the level that Ken demands. The closest thing I've found to a Mullard GZ34 is an RCA 5Y3 I have. Ken points out that it depends on what week the tube was made. One week it might be DuPont chemicals, and the next week Dow chemical coatings. Everything affects tone.

SOLID STATE RECTIFIERS

In many amps you'll see two or three solid state diodes connected in a series string (end to end). This is to increase the Voltage rating of the total diode string. To do this properly, a 1 meg to 10 meg resistor should be placed across each diode. This balances the Voltage equally across each diode.

The modern series of 1N4001 to 1N4007 diodes vary in Voltage rating from 50 to 1000 Volts, at 1 Amp of Current. A 100-Watt tube amp uses a maximum of about 400mA of Current. So, any of these diodes will work. Since they all cost about $.07 each/per hundred, you might as well use the 1000 Volt 1N4007. If you need a higher Voltage diode, the 1500 Volt ones cost about $4.50 each. I use the BY133, rated at 1300 Volts, at 1 Amp, and a thousand of them cost me just $.04 each! In solid state amps I use a 1N5399 rated at 1000 Volts, at 1.5 Amps, or the 1N5408 rated at 1000 Volts, and 3 Amps of Current.

BOMB PROOFING RECTIFIER TUBES

So what happens if a rectifier tube shorts out? You'll blow the fuse, and maybe the power transformer. But there's an easy way to protect the tube without sacrificing its sound. Add solid state diodes to the high Voltage AC wires going into the tube (the red wires on pins 4 and 6). The white stripes should face the tube socket, and please shrink wrap the exposed diode connections. Now, if the tube shorts out, you'll fall back on the solid state diodes, preventing a blown fuse, and possible transformer damage. Of course, you

won't know if your rectifier tube is shorted out, unless you can hear the difference between a solid state and tube rectifier! Remember, there are two diodes inside the tube, and they probably won't both short out. So, you might be running on half a tube rectifier for some time without knowing it. In a professional situation the reliability factor becomes important, and this is good insurance.

RECTIFIER CONFIGURATIONS

The four most common rectifier configurations; Half-Wave; Full-Wave Center Tap; Full-Wave Bridge; and Dual Full-Wave Center Tap.

The Half-Wave uses one diode and produces 60 cycle hum. The Full-Wave uses two diodes and produces 120 cycle hum.

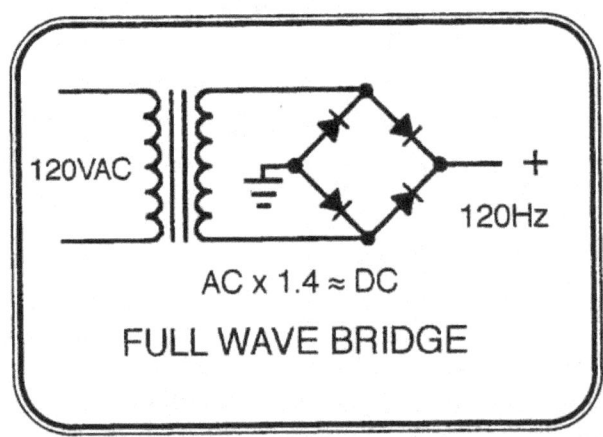

The Full-Wave Bridge uses four diodes, is more efficient, requires a lower AC Voltage in, but at higher Current, and also produces 120 cycle hum.

The chart on the following page, compares the performance of the three most common rectifier circuits used in tube amps.

	HALF WAVE	FULL-WAVE CENTER TAP	FULL WAVE BRIDGE
Output Hertz	60	120	120
Avg. Current	1.00	0.50	0.50
Peak Current	3.14	3.14	3.14
RMS Current	1.57	0.785	0.785
Inverse Voltage	3.14	3.14	1.57
RMS Ripple	121%	48.2%	48.2%
Efficiency	40.6	81.2	81.2

You can see that the Full-Wave Center Tap configuration is better than Half-Wave, in that it puts less stress on the diodes, has higher efficiency, and produces an easier to filter 120 cycle hum. Remember, capacitors like to pass high frequencies. A filter capacitor will conduct 120 cycle hum to ground easier than it will 60 cycle. This results in more filtration with less capacitance.

VOLTAGE, CURRENT & POWER VALUES

When we talk about transformers and power supplies, we refer to Peak Voltage, RMS Voltage, and VA. You might be wondering exactly what these terms mean. Peak Voltage is the highest Voltage reached during any part of the AC cycle. It is measured from zero Volts to the top of one half of the wave. Peak to Peak Voltage is measured from the bottom of a half wave to the top of the other half wave.

VA is Volts multiplied by Amps. It is a true power rating for an inductance. It is sometimes called VARS, or Reactive-Volts-Amperes. Now, Volts times Amps is normally called Watts, but transformer companies don't talk in Watts because it doesn't accurately reflect the demands made on a transformer. VA is a truer rating, taking into account the ratio of resistance to inductance in the device. This can be explained by saying that the Voltage and Current sine waves are not in phase, and the phase lag depends on this inductance/resistance ratio. Therefore, the values you multiply are not peaking at the same time, resulting in an inaccurate value. VARS takes the phase angles into account.

Now, for the most common measurement. RMS is the average Voltage of an AC sine wave. But, how do you average a sine wave? The average value of a sine wave always means the half cycle average. This is determined by drawing the wave, and dividing its time base into ten equal parts. A vertical line is drawn at each division, and its length measured. These lengths are then added together and divided by ten, resulting in the average Voltage value. This type of average is called a mean average. If you perform this operation you'll find that the average Voltage of a sine wave is approximately .636 times the peak Voltage.

RMS means Root Mean Squared. The RMS value is obtained by taking the ten individual measurements and squaring them. The squared numbers are then mean averaged, and the square root of the average is taken. So, RMS is backwards of the actual operation. It's the Square of the values, Meaned, and Rooted. This results in an average of .707 the peak Voltage. The reason this is important, is it more closely relates the heating characteristics of an AC sine wave to an equivalent DC Voltage.

THE HALF-WAVE RECTIFIER

The half-wave rectifier uses one diode to

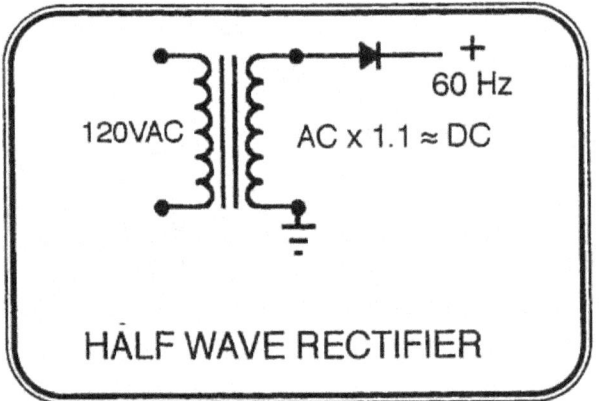

HALF WAVE RECTIFIER

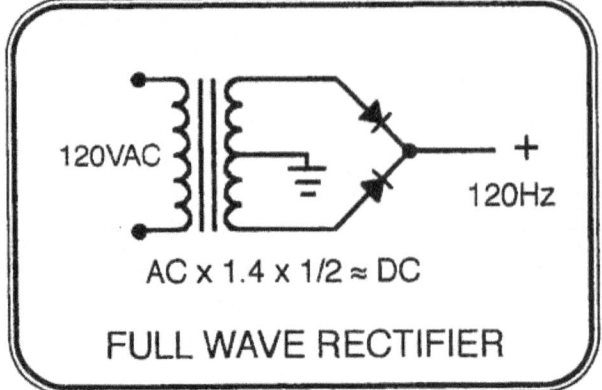

FULL WAVE RECTIFIER

change 60 cycle AC, to 60 cycle pulsed DC. It usually appears in solid state circuits. Since vacuum tube rectifiers usually contain two diodes, most tube amps use full-wave rectifiers, although the bias rectifier in a tube amp is usually a solid state half-wave rectifier.

In Fender tweed amps, you'll find a stack of square steel plates attached to the chassis. This is a selenium rectifier, an early solid state device. It is used to create the bias Voltage, and can be replaced with a modern silicon diode, if it blows.

The single diode rectifier is connected to one side of the transformer's secondary winding, while the other side of the winding is grounded. It's efficiency is low because it only rectifies one half of the AC sine wave (a half-wave rectifier). Without the grounded winding, no Voltage will appear at the diode.

FULL-WAVE CENTER TAP RECTIFIER

The Full-Wave Center Tap Rectifier uses two diodes to turn 60 cycle AC into 120 cycle pulsed DC. It requires a center tapped transformer, with the center tap connected to ground. It has twice the efficiency of the half-wave since it rectifies

both sides of the AC cycle. This is the most common rectifier used in tube amps. The hum it produces is 120 cycle. The 60 cycle hum you hear in an amp is mostly 60 cycle hum from the 6.3 Volt filament supply. If you don't ground the center tap, you won't have any DC Voltage developed as the two cycle will gyrate around each other, canceling each other out.

FULL-WAVE BRIDGE RECTIFIER

The Full-Wave Bridge rectifier doesn't need a center tap, as it artificially creates its own. It is almost never used in tube amps, as the common tube rectifiers have a common Cathode. In order to implement this configuration with a tube, you would need a tube with totally separated Cathodes. They're available, but relatively hard to get.

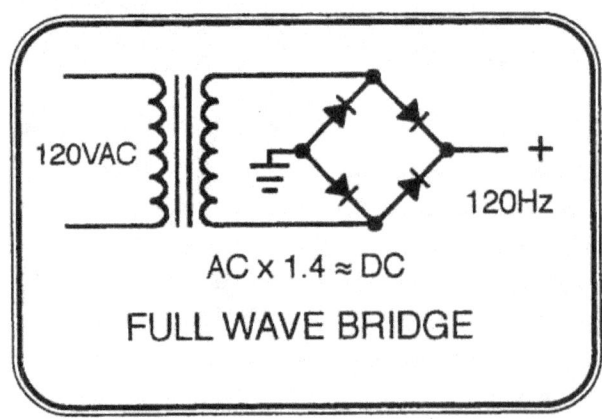

FULL WAVE BRIDGE

I recently put a tube rectifier in a Vox amp with a solid state bridge rectifier. This was done by using the tube for the front two diodes, and leaving the back two solid state diodes in place. The front two solid state diodes had their Cathodes separated, but left in place for short circuit protection, if the tube rectifier fails

DUAL FULL-WAVE CENTER TAP RECTIFIER

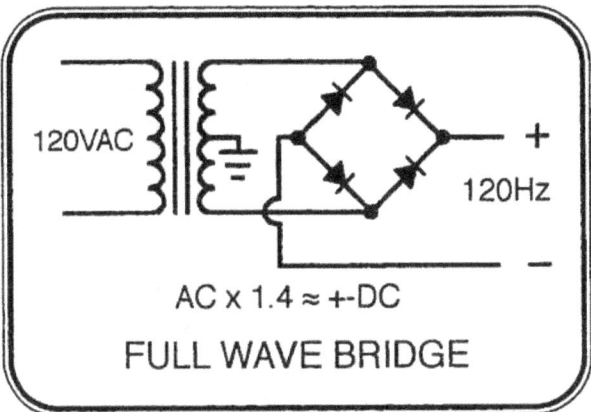

120VAC

120Hz

AC x 1.4 ≈ +-DC

FULL WAVE BRIDGE

The Dual Full-Wave Center Tap rectifier is used to produce both a positive and a negative Voltage. Operational Amplifiers (opamps) usually require what is known as a "split" supply. Notice that the DFWCT rectifier again needs a center tapped transformer.

VOLTAGE DOUBLERS

The voltage out of a rectifier circuit can be doubled by using a circuit known as a Voltage Doubler. There are also Voltage Triplers, Voltage Quadruplers, etc. You can add as many of these Voltage increasing stages as you want, but the regulation deteriorates with each stage.

POWER SUPPLY REGULATION

The regulation percentage of a power supply is the No-Load Voltage, minus the Full Load Voltage, divided by the No-Load Voltage, multiplied by 100. It is basically the transformer's "sag."

If a supply puts out 500 Volts under no-load, and 450 Volts under full load, the supply would be said to have a regulation of 10%. This is a normal performance specification for most tube amps that use "EI" transformers.

TRIODES

As an experiment to control the flow of Current through the newly discovered vacuum tube diode, a third element, called the Control Grid, was added by winding fine wire on two support rods, extending the length of the Cathode. The spaces between the turns are comparatively large so as to not block the passage of electrons from the Cathode to the Plate. There's a special version of Pentodes, and Tetrodes, called Remote Cutoff, or Variable MU, and these tubes have Grids wound tighter at the ends, and looser in the middle. These tubes are designed to reduce modulation-distortion and cross-modulation in radio-frequency stages.

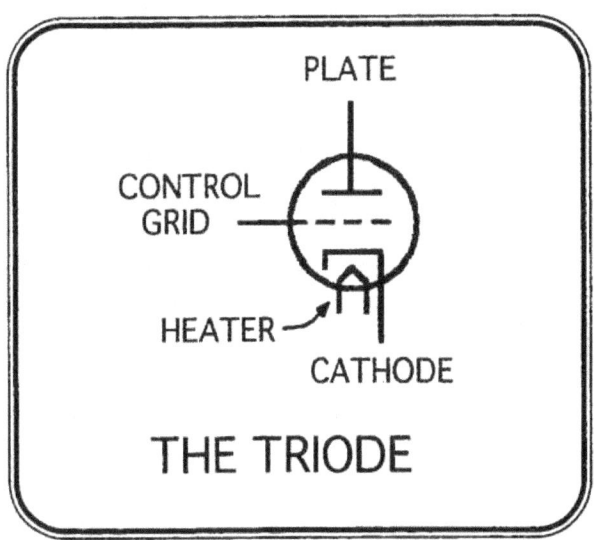

PLATE

CONTROL GRID

HEATER

CATHODE

THE TRIODE

This new Control Grid (abbreviated G1) can interfere with, or even stop, the flow of electrons through the tube. To do this, the Grid must be Negative in relation to the Cathode. This Negative Voltage is called the "Bias" Voltage. The Bias Voltage is set to a point Negative enough so the grid can electrically "hide" the Positive Plate from the Cathode. The Bias Voltage turns the tube off.

When a signal is applied to the Grid, the Positive going signal cancels some of the Negative Bias Voltage, and the tube will start to turn on, and conduct Current. The higher the signal Voltage, the more the tube will turn on. At some point, no increase in Signal Voltage can turn the tube on any more.

This limitation is caused by a number of factors. Two that are related are the Plate Voltage, and the Plate Current. The fact that Current will flow through a vacuum tube, yet the tube does not create a short circuit ("infinitely" high Current flow due to zero resistance) shows that there must be some resistance inside the tube. There are two types of tube resistance, DC Plate Resistance, and AC Plate Resistance.

DC PLATE RESISTANCE

For a certain amount of Plate Voltage, a certain amount of Plate Current will flow. As the Plate Voltage is increased, more Current will flow. These results can be plotted on a chart, called the EpIp curve. By dividing the Plate Voltage by the Plate Current, the DC Plate Resistance is found. This is a normal ohms law solution. If we were measuring a resistor, a steady increase in Voltage, would result in a steady increase in Current. This is the

definition of a linear device. Tubes, however, are non-linear. As the Plate Voltage increases, the Plate Resistance decreases, resulting in a non-linear (more rapid) increase in Plate Current. A value of DC Plate Resistance then, is only accurate for one value of Plate Voltage. DC Plate Resistance is notated as Rp.

AC PLATE RESISTANCE

Tubes are commonly used to amplify AC signal Voltages, which are similar to fluctuating DC. When a signal passes through a tube, it fluctuates the DC Voltage on the tube's Plate, causing an ever changing Plate Resistance to occur.

The AC Plate Resistance is notated rp to distinguish it from the Rp of DC Plate Resistance. It is equal to the ratio of a small change (delta) in Plate Voltage to a corresponding change in Plate Current. The delta represents the fluctuating signal Voltage, and the smaller the delta used in the measurement, the more accurate the results will be. The AC Plate Resistance (rp) can be one-half, or less of the DC Plate Resistance (Rp). When the term Plate Resistance is used, it normally means AC Plate Resistance unless otherwise specified.

TETRODES

The Cathode, and the Plate are solid Plates of metal, formed into tubes. The filament is at the center, with the Cathode tube surrounding it. The control grid is a spiral of wire wrapped around a pair of insulated posts, surrounding the Cathode. The Plate tube surrounds the control grid, and forms the part of the vacuum tube visible from the outside. The whole assembly is attached to mica washers at the top and bottom, keeping everything in alignment, and maintaining the proper spacing.

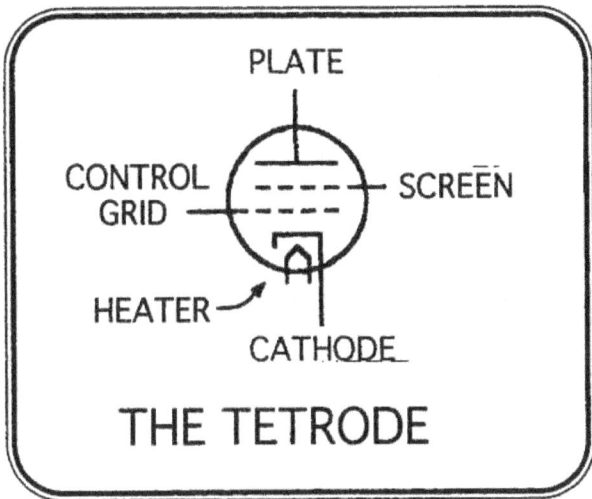

PLATE

CONTROL GRID — + SCREEN

HEATER

CATHODE

THE TETRODE

The fact that we have conductive elements separated by an "air" gap (vacuum gap), creates capacitors between the heater, and the Cathode; the Cathode, and the Grid; and the Grid and the Plate. The amount of capacitance is only a few puffs, which doesn't matter much in audio, when you consider that you have 800 puffs in a standard 20 foot guitar cord, but it is significant in radio frequency work, a forte of early vacuum tubes. The capacitor formed, allows some of the AC signal on the Plate, to couple back to the Grid. Since it's coming off the Plate, it's out of phase, you have a little Negative Feedback Loop, which cancels the stages Gain. Since Capacitors have less Impedance at high frequencies, the result is a loss of High Frequency Gain.

In order to cut down on this Capacitance, a second Grid, called the Screen Grid (G2), was added between the Control Grid and the Plate. Tubes with Screen Grids are called Tetrodes, because they contain a total of four elements. The

Screen Grid divides the Capacitor formed by the Control Grid and the Plate, in half. These two capacitors are then in series. Series Capacitors divide their Capacitance in half (if they have equal values). This makes for a smaller total capacitance between the Cathode and the Plate. Add a Decoupling Capacitor to the Screen Grid, and you've further reduced the Inter-electrode Capacitance of a Tetrode, extending its high-frequency range.

This additional Grid has other effects, as well. As long as the Plate Voltage is higher than the Screen Voltage, the Plate Current in a Tetrode depends to a great degree on the Screen Voltage, and very little on the Plate Voltage. The reason is, the Screen Grid is connected to a high positive Voltage, usually slightly less than the Plate Voltage; is physically much closer to the Cathode than the Plate is; and compared to the Plate, its DC Voltage is practically constant because of the relatively small screen Current flow. This reduces non-linearities in the tube, allowing a higher Amplification Factor.

Because the Screen is an open mesh of wire, most of the electrons flowing from the Cathode to the Plate, pass right through it. Some of the Current does get attracted to the Screen, and flows through it externally, back to the Cathode. The purpose of the Screen Resistor, is to limit this flow of Current to within the dissipation (Wattage) rating of the screen.

A third effect of the screen grid is it puts a hair trigger on the control grid, acting as a "super-charger." Its positive Voltage helps to overcome the control grid's negative bias Voltage, making the tube more sensitive to the input signal. Sort of an avalanche effect, giving the grid more leverage. This gives tetrodes more gain than triodes.

If the Screen happens to be attached to a higher Voltage than the Plate, it abnormally increases the amount of Screen Current. The electrons are more attracted to the Screen's higher Voltage, than they are to the Plate.

Some tweed amps have the Screens connected directly to the Output Transformer's center tap, without Screen Resistors. Because the Plate Voltage has to flow through the Output Transformer's Primary windings, the Plates end up with less Voltage on them than the Screens. I don't find this to be part of the "tweed" sound. The Output Transformer was designed with this in mind, and with Beam Power Tubes, it's not really a problem. Add Screen Resistors, unless you don't like the change in sound.

Examples of Beam Tetrodes are the Mullard KT66, KT77, and KT88 (Kinkless Tetrodes).

PENTODES

A problem exists with Tetrodes, called Negative Transconductance. If the Screen Grid has a Voltage higher than the Plate, Current will flow from the Cathode to the Screen, instead of to the Plate. If the Plate Voltage is increased, some Current will be attracted past the Screen to the Plate; bounce off; and then be attracted back to the Screen's Voltage, which is still higher than the Plate's. This is called "Secondary Emission." The result is that as the Plate Voltage gets higher, its Current increases for awhile. Then it actu-

ally starts to decrease, as the stream of electron Current is stolen by the Screen. This Secondary Emission adds to the Screen's Current, causing it to increase, and resulting in "Negative Transconductance."

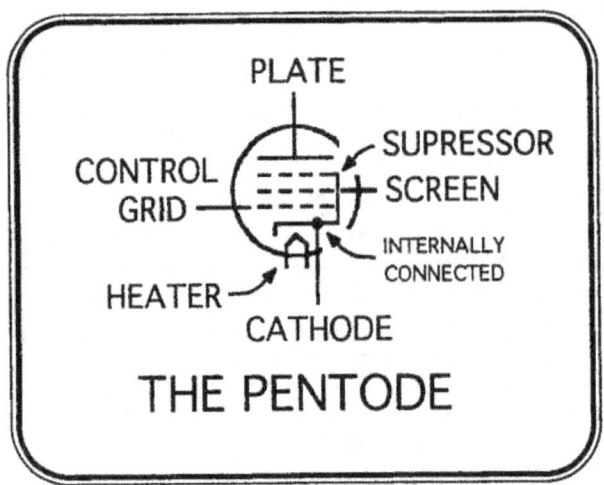

THE PENTODE

A fifth element, the Suppressor Grid, was added to counteract this phenomenon. Normally, the Suppressor Grid is permanently connected to the Cathode inside of the tube. As the electrons bounce off the Plate, and are attracted back to the Screen; the Suppressor repels them back to the Plate.

In Power Pentodes, the Suppressor Grid allows higher power output from a lower Grid Voltage. This results from the fact that the Plate Voltage swing can be made very large. In fact, the Plate Voltage may be as low as, or lower than, the Screen Grid Voltage without serious loss in Signal Gain capability.

The EL34 does not have its Suppressor connected inside the tube. Rather, it is brought out to pin 1, and this needs to be connected to the Cathode (pin 8), externally. Be aware that Fender used pin 1 of the tube socket as a wiring terminal. This didn't matter with 6L6's, since the 6L6 doesn't use pin 1. If you want to substitute EL34's into a Fender amp, you'll have to disconnect the signal wire, and the grid resistor from pin 1; re-connect the wire, and resistor together; shrink wrap them; and then short pin 1, and pin 8 together.

Examples of Pentode tubes are the European EL34, and EL84, and the 6267 preamp tube used in Vox AC15's.

BEAM POWER TUBES

The 6V6, 6L6, 6550, and the American EL34, and EL84 are not really pentodes. Instead of a suppressor grid between the screen, and the Plate, beam power tubes actually have beam-forming Plates on opposite sides of the Plate.

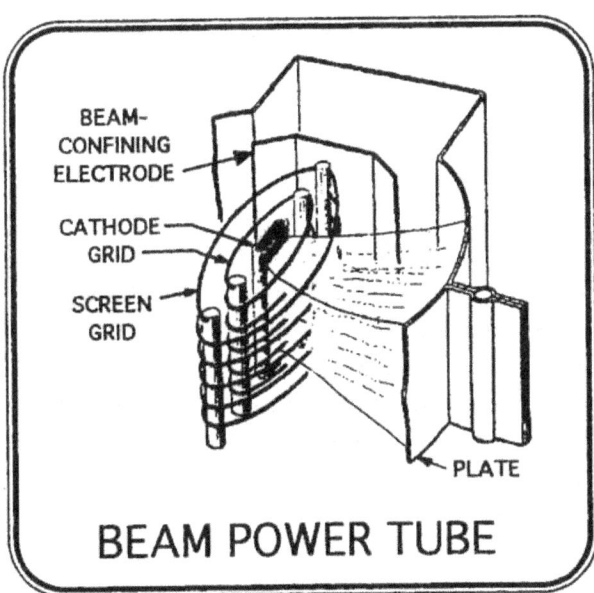

BEAM POWER TUBE

These Plates "focus" the beam of electrons onto the Plate by forming an electrostatic field. In beam tubes, the control, and screen grid windings are set directly in line with each other, with respect to the Plate. The total effect is to inhibit secondary emission. In effect, beam power tubes operate as pentodes, but are actually Beam Tetrodes, and this distinction is of little concern. The tubes that start with KT, as in KT66, and KT88, are Kinkless Tetrodes, another name for a Beam Tetrode.

TRIODE/TETRODE/PENTODE SWITCHING

Amps which have "triode" settings on them, cannot actually operate as triodes. These switches disconnect the screen grids, or connect them to the Plate, but the beam-forming Plates (suppressors) are internally connected, and cannot be disconnected, except on the EL34. And, even if they are disconnected, or tied high, they're still physically there, and affect the sound.

PENTAGRIDS/HEPTODES AND OTHER SPECIALTY TUBES

Some tubes contain two or three triodes in a single package. Some tubes even have a triode, and a pentode in the same package. Single tube sections that go beyond the pentode, and contain more than five elements, are called multielement tubes, and can be made to act as more than one tube "interleaved" together. If you have more than one control grid inside a tube, you can "mix" two signals together, as in superimposing an audio signal onto a radio frequency carrier. The heptode, also called the pentagrid, is and example of this. These tubes are found in some vacuum tube compressors, and perform the double function of oscillator and mixer in superheterodyne receivers.

71

CHAPTER 9
HOW TUBES REALLY WORK

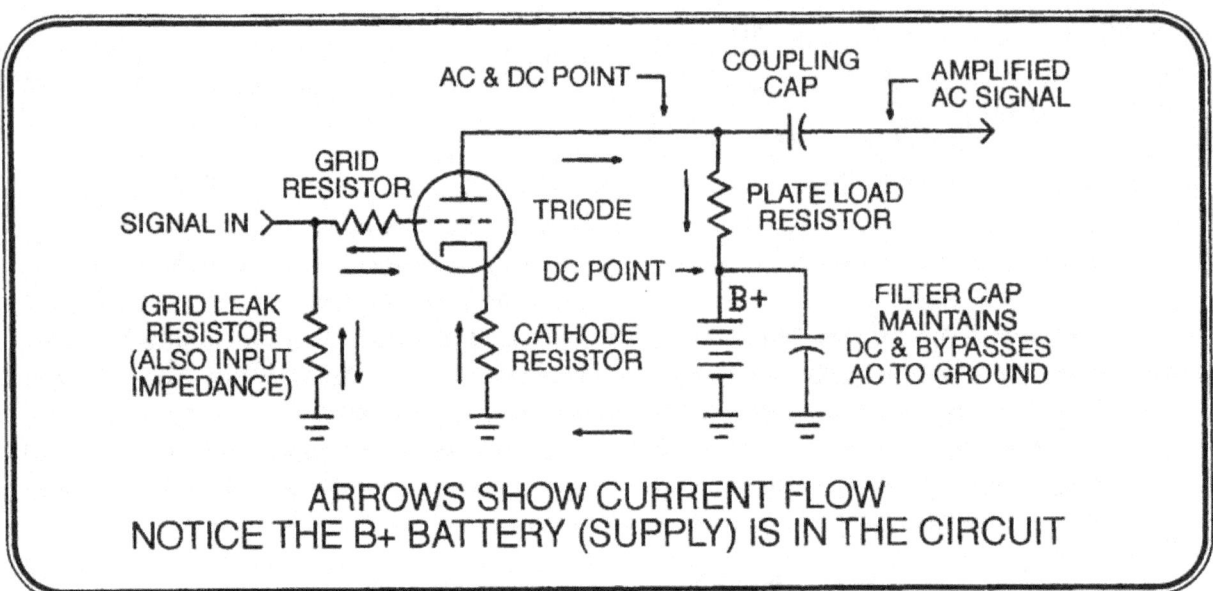

AC & DC POINT

COUPLING CAP

AMPLIFIED AC SIGNAL

GRID RESISTOR

SIGNAL IN

TRIODE

PLATE LOAD RESISTOR

DC POINT

B+

GRID LEAK RESISTOR (ALSO INPUT IMPEDANCE)

CATHODE RESISTOR

FILTER CAP MAINTAINS DC & BYPASSES AC TO GROUND

ARROWS SHOW CURRENT FLOW
NOTICE THE B+ BATTERY (SUPPLY) IS IN THE CIRCUIT

HOW TUBES REALLY WORK

In this chapter, you'll learn how the different elements of a tube operate, and what the tube's specifications mean.

If you add an audio signal to the control grid, it gyrates the Grid Voltage around the Bias Voltage, either turning the tube more on, or more off. Since the resultant Plate Current is large compared to the Input Signal Voltage, the effect is that a small Signal Voltage on the Grid controls a large Current. The large Current through the Plate Resistor creates a large Output Voltage, resulting in amplification. In this way, a small signal Voltage controls a large output Voltage.

If the tube loses its Bias Voltage, the tube will turn full on, overheat, and turn cherry red. It's the equivalent of getting your gas pedal stuck to the floor.

THE GRID LEAK RESISTOR

In much the same way, only different, is that without the Grid Leak Resistor, the Grid would absorb some of the electrons flowing from the Cathode to the Plate, and eventually become excessively negative, turning the tube off. The Grid Leak Resistor gives those electrons a path to ground, hence its name, "Grid Leak." In doing so, it also references the grid to ground.

LOAD RESISTANCE

The value of Load Resistance controls the linearity, and to an extent the gain of a tube stage. The larger the load resistor, the higher the gain, and the more linear the stage will become. This assumes adequate values of grid bias, and Plate Voltage. Higher Plate Voltages, allow higher values of load resistance, and also require higher bias Voltages (more negative). Higher bias Voltages allow the tube to accept larger input signals.

Since the output signal Voltage is directly proportional to the Plate Current, a linear relationship between the Grid Voltage, and the Plate Current, will produce a linear relationship between the input and output signals.

Since the Plate Current multiplied by the load resistance equals the Output Voltage, at first you might think that the lower values of Plate Current would produce a smaller Output Voltage. In fact, the higher resistance more than makes up for the small Plate Current.

TIP: You can modify the distortion characteristics of a preamp stage by changing the value of the Plate Resistor. Common values are 47K through 330K, 100K being the norm, and 220K being the second most common. Usually the second or third gain stage in a preamp distortion circuit will have a 220K resistor. You might think of it as changing the flavor of the second stage, so instead of just boosting the gain, you get a different type of distortion tone mixed in. Here we get subjective. Don't ask me what works best. That's what your ears are for. Try it. And please, don't go around changing the Plate Resistors on every Fender amp to 220K. I'll just be changing them back to 100K. I've seen some very good sounding modified Marshalls that had all 100K Plate Resistors.

The best way to do this type of Plate Resistor tuning is to install a 47K resistor, with a 250K pot wired as a variable resistor. You can then quickly dial between 47K and 297K. When you find the sound

you like, unplug the amp; make sure the B+ Voltage is discharged; and measure the total resistance across both the 47K resistor, and the pot. You've now found the perfect Plate Resistor value. Try doing this with ever resistor value in the amp, and remember that everytime you change one value, it affects every other value in the amp, and you'll see why amp building is an empirical science.

PLATE VOLTAGE

The Plate Voltage serves two functions. It must be large enough to provide the desired output Voltage, and have enough left over to properly operate the tube. If the value of the Plate Resistor is made too high, so much of the Plate Voltage will be dropped across it that insufficient Voltage will be left to operate the tube, and distortion will result.

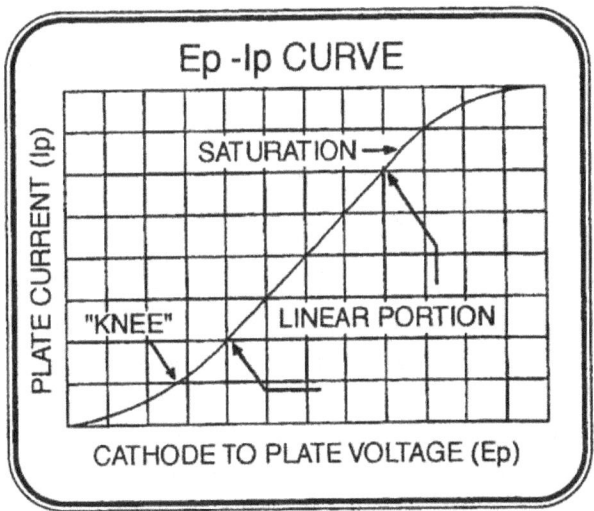

Too low a Plate Voltage will provide insufficient gain and output distortion due to improper tube operation. There's a distortion foot pedal that advertises "real tube distortion" that in fact operates its tube at such a low Voltage, its barely on. The pedal actually uses an opamp with a diode clipper to create the distortion that is then passed through a tube as a sort of "filter" to slightly soften the solid state buzz. If you want to design a distortion pedal using tubes, I would suggest an absolute minimum of 100 Volts on the tube's Plates.

Regardless of the Plate Voltage, if the maximum Plate Current is exceeded by the circuit's design (bias Voltage, and/or drive levels, or load impedance), the tube will overheat, and melt. If someone tells you that you can get more power out of a tube than it was designed for, because of some design trick, you should permanently mount a fire extinguisher to the side of your amp.

AMPLIFICATION FACTOR

Amplification Factor is the ratio of Grid Voltage effectiveness, over Plate Voltage effectiveness in producing Plate Current. An Amplification Factor of 100 means, a 100 Volt increase in Plate Voltage would be required to cause as much of an increase in Plate Current, as a one Volt increase in Grid Voltage would. In other words, the grid is 100 times more effective in controlling the flow of Current through the tube, as the Plate Voltage is.

Amplification Factor is notated by the Greek letter "μ," and is pronounced "mu." Technically it's defined as the small change (delta) in Plate Voltage divided by the small change in Grid Voltage, that would produce the same change in Plate Current.

$$\mu = \frac{e_p}{e_g} = \frac{\text{small change in Plate Voltage}}{\text{small change in Grid Voltage}}$$

Triodes that have mu's of less than 10 are called low-mu triodes. Triodes with mu's between 10 and 30 are called medium-mu triodes. Triodes with mu's above 30 are called high-mu triodes.

Common amplification factors for preamp tubes are:

12AX7	=	100
5751	=	70
12AT7	=	60
12AY7	=	40
12AU7	=	17

The Amplification Factor doesn't mean the Voltage Gain of the tube. It is only a measure of how much better the Grid Voltage is at controlling Plate Current, than the Plate Voltage is.

The Plate and Grid Voltages control the Output Voltage by varying the Plate Current. You can change these Voltages by soldering in different part values, but it would be nice to have an easier way to change the Output Voltage without opening up the chassis. An easy way to do this is to simply change to a tube with a different Amplification Factor. Change the inverter/driver tube from a 12AT7 to a 12AX7, and you've automatically increased your gain.

Of course, nothing is that easy. If you increase the gain too much, you could have stability problems, and get a nasty dirty tone, with no clean tone. If you decrease the gain too much, or in the wrong place, the amp could sound dead, not have any breakup, and lack sensitivity. Sometimes an amp will put out more power with a 12AX7 Inverter tube than with a 12AT7. The reason being, the 12AT7 might not be able to provide a high enough Grid Voltage to the Power Tubes. Then the Inverter distorts before the Power Amp runs out of power. This means that the Power Amp is accurately reproducing a distorted Inverter signal, and sounds like it's clipping due to a lack of power.

My suggestion would be to try a change in each stage, and see if you find something you like. Remember, as you change each stage, it affects all the other stages. The variations are the number of stages, taken to the power of the number of different tube types you're trying. On the same subject, to get the "correct" sound out of a 4-10 Bassman, the first tube stage should be a 12AY7.

TRANSCONDUCTANCE OR MUTUAL CONDUCTANCE

Transconductance is probably the most important tube constant, defined as the ratio of a small change Plate Current, and the change in Grid Voltage that produces it. It's also known as Mutual Conductance, not to be confused with the term "mu" above, and is abbreviated "gm."

$$gm = \frac{ip}{eg} = \frac{\text{small change in Plate Current}}{\text{small change in Grid Voltage}}$$

Transconductance is the ratio of a Current to a Voltage, and is known as a Conductance. It is expressed in a unit called the Mho, which is ohms spelled backwards. This unit has now been renamed the Siemens, abbreviated "S." As with Farads, the Mho is too large a unit for practical use, and is usually expressed as micromhos, which is now abbreviated "μS," or micro-Siemens. A "μS" is equal to a Plate Current change of 1 micro-

Ampere (μA) for each 1-Volt change in Grid Voltage. Therefore, a tube with a Transconductance of 2000 μS has a change in Plate Current of 2,000 μA (or 2 milliAmps) for each 1 Volt change in Grid Voltage.

Transconductance is used to compare tubes in similar applications. A tube with a higher Transconductance will produce a higher Current flow for a given input Voltage, and therefore has a larger output Voltage. Remember, Current flow produces the output Voltage by dropping it across the Plate Resistor (IR Drop). More Current through a tube, produces more output Voltage.

Transconductance gets its name from the fact that a Voltage on one tube element (the Grid) controls by "Trans"fer, the flow of Current (Conductance) in another tube element (the Plate).

TUBE CONSTANT RELATIONSHIPS

The tube constants result from the physical geometries of the tube's design. In other words, a tube can be designed to have specific properties by varying the size, shape, and spacing of its internal elements. The three constants described above have the following relationship:

$$\mu = rp \times gm$$

This means that for the same values of Plate and Grid Voltage, the Amplification Factor will equal the AC Plate Resistance multiplied by the Transconductance.

The inverses are: $rp = \mu / gm$
and $gm = \mu / rp$

CHAPTER 10
WHAT IS BIAS?

Oh, that smell! Have you ever had a valuable vintage amp catch fire? You can smell the power tubes burning up, and you can usually hear the metal Plates inside the tubes creaking like an overheated pizza pan. If this happens to you, turn the amp off immediately. This problem comes from a loss of bias, sometimes caused by a shorted power tube (grid to Cathode short).

To find the bad tube, remove the one that's "gone cherry." In an amp with four power tubes, two might go cherry at the same time. Usually, only one tube is bad. Remove one at a time until you find the bad one. If the problem goes away, consider yourself lucky. Sometimes, when a tube goes bad, it takes other parts of the amp with it.

In older amps a loss of bias can be caused by a worn out bias supply filter cap, or a broken wire inside the amp. Then you should take the amp to a qualified service shop for repairs.
But haven't you ever wondered what is bias, and why is it important? Are there audible advantages to adjusting my bias correctly, or incorrectly? Let's start with the way tubes are biased.

METHODS OF BIASING TUBES

It almost a guarantee that a preamp tube will be operated in Class "A." Power tubes are usually biased to run in Class "A," Class "AB," or Class "B." Whatever class the tubes are biased into, there are seven

ways to get them there:

      1. Grid-leak bias
      2. Alternate grid-leak bias
      3. Cathode Bias / Self-Bias
      4. Fixed bias
      5. Battery bias
      6. Contact bias
      7. AGC bias

These methods of bias are sometimes combined producing even more permutations, as in the Fender Silver Face amps which use a combination of Cathode and Fixed bias.

CATHODE BIAS / SELF-BIAS

Cathode Bias is also known as Self-Bias, or Automatic Bias. This is the most common form of bias used in guitar amps. Every preamp stage I can think of is biased with Cathode Bias. Every Class "A" power amp I know of is biased with Cathode bias. It is almost a given, that if the tube is biased into Class "A," it is done with Cathode Bias. The reasons for this will become obvious, as we continue.

Cathode Biased amplifier circuits only require service if the "Cathode Cap" or "Cathode Resistor" fails. This will be noticed by a loss or increase in gain, or a hum in the amp.

The power tubes in a Class-A power amp are also generally Self-Biased, and are adjusted as part of the amp design process. There is really no reason to

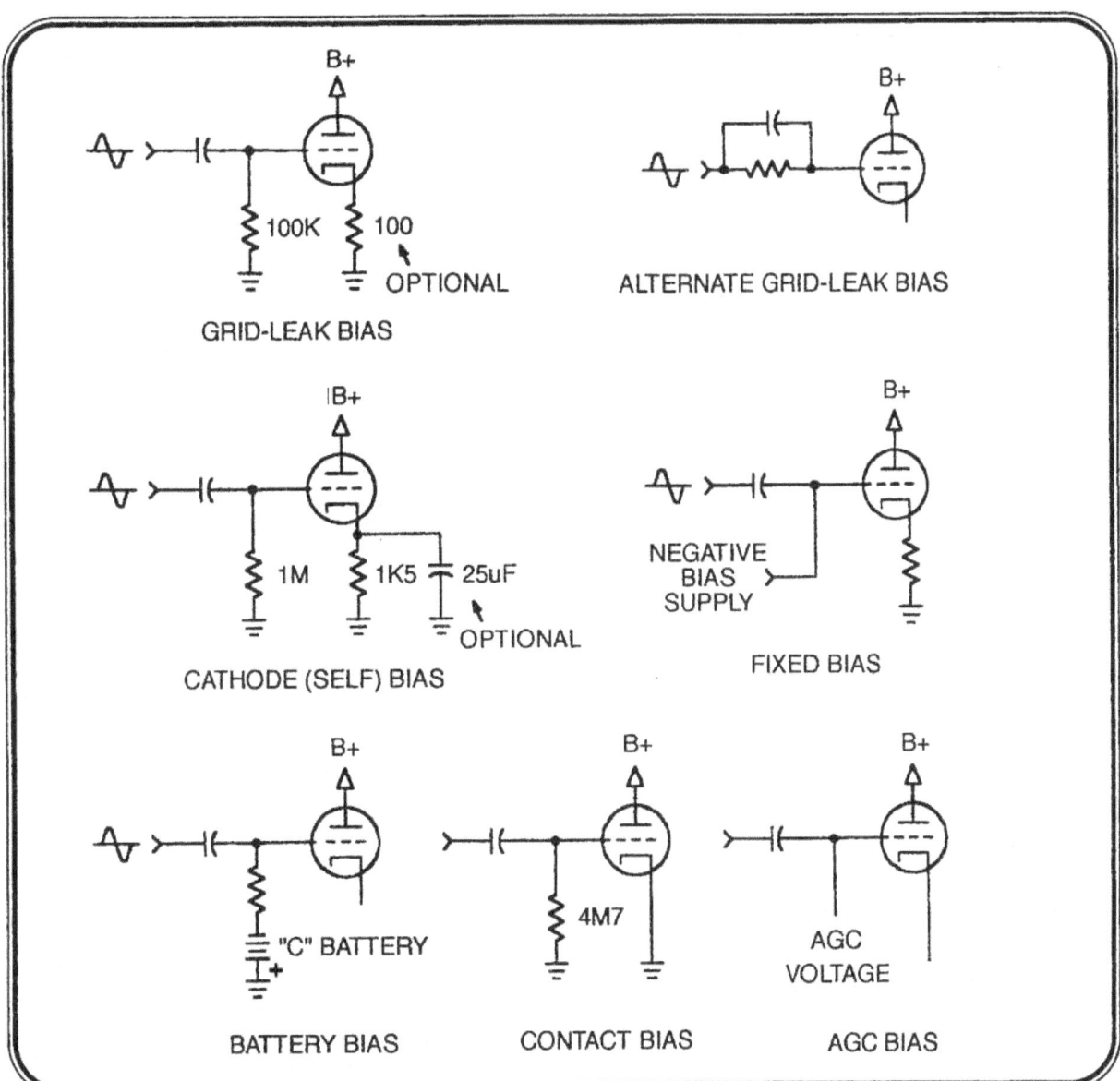

GRID-LEAK BIAS

ALTERNATE GRID-LEAK BIAS

CATHODE (SELF) BIAS

FIXED BIAS

BATTERY BIAS

CONTACT BIAS

AGC BIAS

mess with someone else's design, but slight adjustments are possible. Fender tweed amps, many Gibson amps, and Vox amps are the most common Class-A power amp designs.

WHAT DOES BIAS DO?

We've approached this subject previously, from different directions. A tube needs to have its Grid negative RELATIVE to its Cathode. If the Cathode is positive, and the grid is grounded, then the grid IS neg-

ative RELATIVE to the Cathode. With Cathode Bias this is accomplished by adding a resistor between the Cathode and ground, while connecting the Grid to ground through a Grid-leak resistor. As the tube conducts Current, a Voltage drop (IR Drop) is developed across the Cathode Resistor, making the top of the resistor, and the Cathode, positive.

The Cathode Resistor also serves a Current Limiting function. As the tube conducts more Current, the IR Drop increases, making the Grid relatively MORE negative, helping to turn the tube off. This is why it is sometimes known as Automatic Bias.

Class "A" push-pull operation is covered in its own section, but for now, something to remember about Class "A" Push-Pull is, the amount of Current flowing through the tubes in a properly designed amp, does not vary widely.

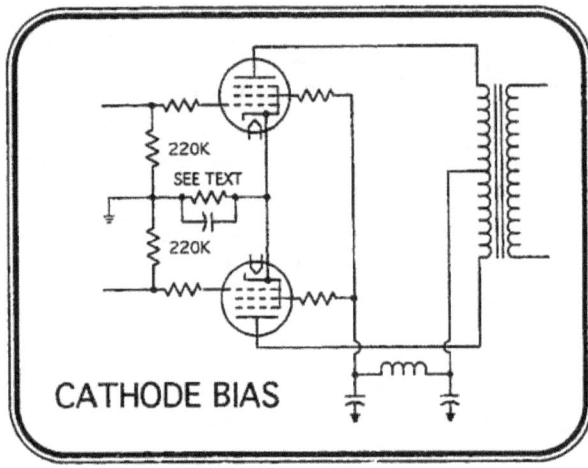

CATHODE BIAS

If the Cathode Resistor is made too large, the IR Drop through it will cause the Voltage on the Cathode to fluctuate widely, causing compression, power loss, and distortion. The mistake being made is using the Cathode Resistor to excessively Limit the Current through the tube in the Idle (zero signal) Mode.

The mistake is in making too little Current flow at idle (20-30mA per tube). As the tubes turn on, more Current flows. Since the bias Voltage is the product of the Cathode Current times the too large Cathode Resistor, the increase in Cathode Current causes the Cathode Voltage to rise excessively, shutting the tube off. This results in a severe loss of power, large amounts of crossover distortion (you though that only occurred with

Class "AB" amps?), and severe compression, which could also be described as a lack of dynamics, or headroom. The way to stabilize the bias Voltage is to stabilize the Cathode Current, and this is not done by using a large Cathode Resistor, although there is an expensive amp out there that advertises that it's a Class-A design, but it's only that with small signals. It's really a Self-Biased Class-AB amp. The Cathode Resistor is more than doubled in size. If this amp sounds loud, it's due to the compression this causes, and its high amount of distortion (see Chapter 22 about pyscho-acoustic effects).

"Changes in effective gain during " low-frequency " transients occur in amplifiers with output stages of the self-biased Class AB type, causing serious distortion which is not revealed by steady-state measurements. The transient causes the current in the output stage to rise, and this is followed, at a rate determined by the time constant of the biasing network, by a rise in bias voltage which alters the effective gain of the amplifier." (The Williamson Amplifier, by D.T.N. Williamson.)

This distortion is heard at the higher harmonic frequencies it creates, and is most likely the cause of the bark, and "sub-harmonic" trailing notes that you hear when you play above the 12th fret through this amp. This is, if I may say, an error, and I'm surprised the magazine reviewers didn't catch this. Maybe they don't play above the 12th fret.

The strange thing is, as the Cathode Resistor is made smaller, the resulting positive Cathode Bias Voltage doesn't go down much. The decrease in the Cathode Resistor value, is offset by the increase in Cathode Idle Current (to around 50-60mA per tube) in the IR Drop formula. Now, as the tubes turn on more, the Bias Voltage is much more stable as the Cathode Current is multiplied by a smaller Cathode Resistor value. Also, since the tube is already running "hot," the Cathode Current doesn't increase much. This makes sense when you remember that the whole point of Class "A" operation is the tubes are biased in the "middle;" half on, and half off, although this is actually the 100 percent dissipation point. Remember, that as the tube turns on, with an increase in signal above the half way point, it also turns itself off on the next half-cycle with a decrease in signal strength. Net result? 100% dissipation, at idle, and at all times for that matter. That's why they run so hot. But trying to turn the tubes off with a large Cathode Resistor just doesn't work well.

SECTION 8 : CLASS AB₂ AMPLIFIERS

(i) Introduction
(ii) Bias and screen stabilized Class AB₂ amplifier
(iii) McIntosh amplifier.

(I) Introduction
    Class AB₂ amplifiers closely resemble Class B₂ amp as for Class AB₁ operation. Consequently they are l ing, and the distortion which occurs in the plate circ AB₁. The plate circuit efficiency is intermediate bet —typical values for triodes are from 40 to 48%. Th zero to maximum signal is less than with Class B oper
    The matching of valves is covered in Sect. 5(v).
    Pentodes and beam power amplifiers may be used operation.
    Fixed bias is essential. ◄──
    Type 6L6 or 807 beam power amplifiers may be use efficiencies of about 65% (or 61% including screen

This is why Self-Biasing tubes into Class "AB" operation is futile. I'm not saying it can't be done. I'm saying that it doesn't work well. I believe you can drive a car with a flat tire. It doesn't mean it's a good idea. The whole point of Class "AB" oper-

ation is to turn the tubes off during idle. So, now we're back to trying to use a large Cathode Resistor, which I've already shown doesn't work.

In the Fender Silver Face amps, Cathode Resistors were combined with additional Negative Fixed Bias applied to the grids. This did not work well either.

FIXED BIAS

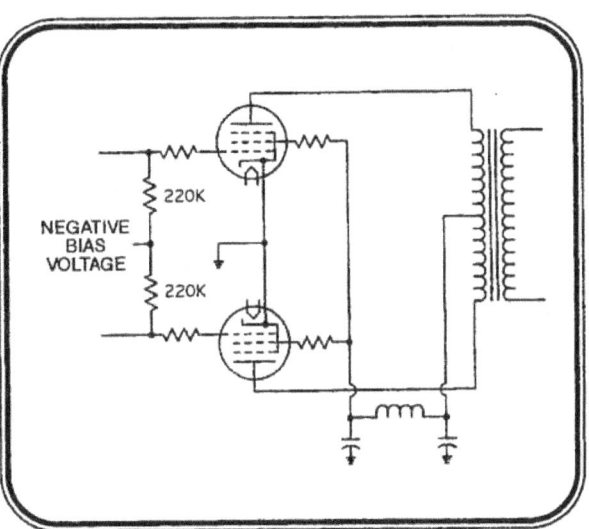

Fixed Bias is developed from a negative power supply, and is applied to the Grid of the tube. It is easily controllable, and stable since only 1 to 2 milliAmps of bias Current is needed, resulting in very small power supply requirements, even when biasing power tubes. The filter capacitors used in the negative Fixed Bias power supply are only 10 to 50uF for this reason. Do not use a cap larger than 100uF for the Bias Supply. Filter capacitors store Current during periods of low demand, to supply it during periods of high demand. In this way, they smooth the Voltage being supplied. Remember, the Grid has very high input impedance because it isn't physically connected to anything inside the tube, and only conducts Current due to:

1. The capacitance formed between the Grid and the Cathode.

2. A resistive element equal to the transit time of electrons between the Cathode and the Grid.

3. A resistive element due to the part of the Cathode lead inductance common to both the input and output circuits.

Components 2 & 3 are dependent on the frequency of the input signal, and result in a very high Input Impedance at audio frequencies, especially when the grid is biased negative. For these reasons, the demand for bias Voltage is very low, and a small capacitor goes a long way in smoothing the bias Voltage. This bias Voltage is very stable, and results in an amp with less BIAS VOLTAGE compression than in a Cathode Biased amp. This should not be misinterpreted to mean that Cathode Biased amps have more compression than Fixed Bias amps. There are other things involved. More on this later.

ADJUSTABLE FIXED BIAS

It sounds like a contradiction, but amps with a Fixed Bias, usually have some way to adjust its fixedness. Later Fender designs are the most common amps with adjustable Fixed Bias. The Bias should be adjusted so that 35 to 40 milliAmps of Current flows through each power tube when no signal is applied. Thirty-six milliAmps seems about right to my ears. The difference in sound can be described

as cold and weak at the 20ma level, or stronger and more solid starting at 25 to 30 milliAmps. Above 40 milliAmps you're asking for short tube life, without much of an increase in the "rawness" of the sound. Notice that starting in 1968, the bias adjustment on Fender Silver Face amps became a bias balance control. These should be modified into a straight bias adjustment control by your repair shop.

FIXED BIAS PREAMP TUBES?

Before you start thinking about using Fixed Bias on the preamp tubes, which could be done, you have to remember that the Cathode Bias commonly used, is easier to implement. If you wanted to put a preamp tube into Class "AB," "B," or "C" operation, you would HAVE to use Fixed Bias for reasons explained above. But preamp tubes are run Class "A" because they are SINGLE-ENDED. A larger negative bias Voltage is needed to push a tube into Class "AB" bias. As shown above, this Voltage CANNOT be created by Cathode Bias. If you were to bias preamp tubes into Class "B" operation, for example, you would need a push-pull circuit, since Class "B" operation only passes 180 degrees of the signal. The push-pull circuit provides two 180 degree outputs which are combined to form the full signal. As I said, single-ended Class "A" preamp tubes are a lot easier to implement, and even sound better!

BATTERY BIAS

Battery Bias is the same as Fixed Bias, except the bias voltage is supplied by a battery. In the early days of radio, all the supplies were batteries. The bias supply battery was called the "C" Battery. You might then call the bias supply the "C+" supply (although nobody does).

CONTACT BIAS

Contact bias is sometimes used with high-mu tubes. Electrons flowing from the Cathode to the Plate, pass through the Grid wires. The electrons that strike the Grid wires need a return path to the Cathode. If a very large Grid-leak Resistor is used, this small amount of current will develop an appreciable amount of voltage across it. Contact Bias can be identified by the large value of Grid-leak Resistor. It needs a high-mu tube since high gain implies relatively small input signal, which require low bias voltage values, and the relatively large Plate Current generated by a high-mu tube means more electrons striking the grid. Two features that help each other.

AGC BIAS

AGC, or Automatic Gain Control Bias, is produced by DC rectifying the input signal to produce a negative DC bias voltage. The larger the input signal, the larger the bias voltage, which tends to reduce the gain of the stage. In this way, the gain of the stage is automatically controlled. AGC circuits are common in video and audio cassette recorders that do not have manual controls for recording levels. This is why you can just plug into a recorder and get a proper level.

CHAPTER 11
POWER AMPLIFIER CLASS

CLASS-A AMPLIFIERS

There's more misunderstanding concerning Class-A amplifiers, than any other subject I could pick. So, let's have at it.

The Maximum Plate Dissipation in a Class-A amplifier occurs at the Zero Signal point. That's the "Idle" point where the amp is at operating voltage, but has no signal running through it.

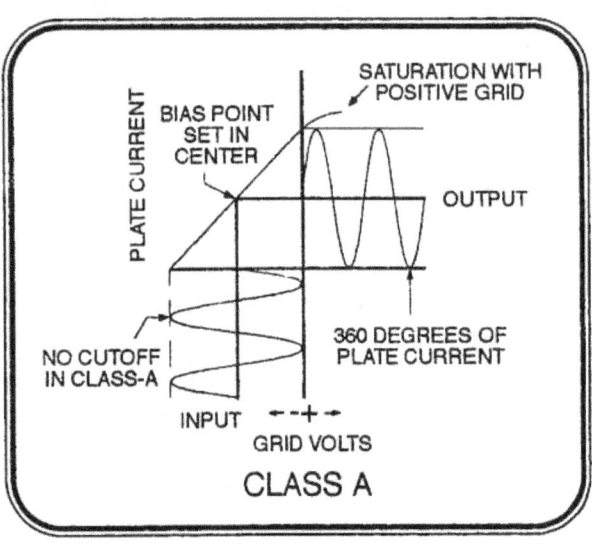

The easiest way to get there is to just turn the Volume control down to "zero." The Plate Dissipation can be calculated by taking the Plate Voltage, multiplied by the Plate Current. The Plate Dissipation for a 6V6 is 12 watts. A 5E1 Tweed Champ runs at 305 Plate Volts, with a 470 ohm Cathode Resistor, and 19 Volts expected on the Cathode.

$$19 \text{ volts} / 470\Omega = .040 \text{ amps or } 40\text{mA}$$
$$305V \times .040A = 12.2 \text{ watts}$$

Right on the money. You can use this type of calculation to determine the proper Class-A biasing for any amp. Remember, the tube will be biased at its full power dissipation AT IDLE. To understand this, remember that the input signal will drive the tube into higher power, AND THEN into lower power. The average power will remain at its maximum.

CLASS-A EFFICIENCY

Class-A has a reputation for lower efficiency, and higher heat, but it also has lower distortion, and usually uses lower Plate Voltages.

Those who don't understand tubes often say that Class-A designs produce less power from a set of tubes BECAUSE Class-A is less efficient. That's wrong. Efficiency relates to how a Class-A design converts wall power into audio power. A Class-AB design might pull 1 Amp of idle current out of the wall. The same design run in Class-A might draw 1.8 Amps. This increased power consumption is the inefficiency referred to, and is the reason that Class-A designs run so hot.

It's not that the tube itself has somehow become less efficient. The lower power output is related to the fact that Class A has lower bias voltages, and in order to keep the tube within the dissipation rating of its Plate, lower B+ voltages are required as well. This results in less power out of the amp circuit. Ultimately, the amount of power you can get out of a tube is a function of its Plate Voltage.

A Class-AB design has a higher (more negative) Bias Voltage, which reduces the current flowing through the tube for the same Plate Voltage. This allows a higher Plate Voltage to be used, without exceeding the Maximum Plate Dissipation of the tube. It's this higher Plate Voltage that allows Class-AB designs to produce higher power output. So, it's not that Class-A designs put out less power. It's that Class-AB designs put out more.

This higher Bias Voltage in Class-AB, also allows a larger drive signal to be applied to the Grid of the power tube. This larger input signal, multiplied by the same Gain, results in a larger output signal. This higher output voltage, which is allowed by the higher Plate Voltage used, again, results in higher output power in a Class-AB design.

CLASS-A COMPRESSION

There's also a belief that Class-A has more compression than AB. I remind you that the entire preamp is Class-A, and doesn't suffer from compression; that the hi-fi market prizes Class-A designs over Class-AB designs; and that a Vox, when played clean, is more dynamic than a Marshall. This idea got started by someone thinking that because Class-A amps put heavy demands on the power supply, and therefore, the power supply MUST have more sag in it, and this sag CAUSES compression. But that's exactly why Class-A has LESS compression. The supply is already "pre-sagged" at idle, which produces a solid foundation to build a signal on. It doesn't sag much more

when a signal goes through, since it's already running full tilt at idle. A Class-AB amp has the supply bouncing all over the place, trying to track the signal, and compressing it as it goes. Anyone who has watched the Plate Voltage change on both types of amps will notice the small amount of change from idle to full power in a Class-A amp.

SINGLE-ENDED OR PUSH-PULL

All amps that use one power tube, like the Champ, are called "Single-Ended," and by definition are Class-A. Class-A means that the tube conducts the entire 360 degrees of the signal. It should be obvious that if there's only one tube, it HAS to conduct the entire signal, since it has no complement to share the duty with.

With Class-A amplifiers, reduced distortion and improved power performance are obtained by using two tubes in a Push-Pull design, even though it requires twice the Grid-Signal Voltage. Distortion caused by Even-Order Harmonics and Hum caused by Plate Voltage supply fluctuations are either eliminated or drastically reduced through Phase Cancellation. Because Distortion for Push-Pull operation is less than for Single-Tube operation, more than twice the Single-Tube output can be obtained.

For either Parallel or Push-Pull Class-A operation of two tubes, all Currents are doubled while all DC Voltages remain the same as for Single-Tube operation. If a Cathode Resistor is used, its value should be about one-half of that for a single tube.

Understand that when a Push Pull stage is running Class-A, both tubes are conducting at the same time, just in opposition to each other. It's only when a Push Pull pair gets pushed into Class-B operation, that the top tube conducts the top half of the signal while the bottom tube "rests." If you biased a Champ into Class-AB, with only one 6V6 tube, a good chunk of the signal would be missing. The single 6V6 tube has to carry the entire signal.

"Can I pull one tube out of my Bassman, and get 25 watts? NO! You'd get 12 watts. Remember, Small Signal Class-AB operation conducts 360 degrees of the signal (Class-A). Large Signal operation conducts less than 360 degrees of the signal (Class-B). It would be like running on a flat tire, with almost half the signal missing. In a Class-A Push-Pull amp, like the Vox, the top tubes are pushing up, while the bottom tubes are pushing down, with all the tubes, conducting all of the signal, all of the time. You might call it a Push-Push, Pull-Pull amp.

Never remove tubes from a Cathode Biased power amp, thinking you're going to turn a Vox AC-30 into an AC-15. The Bias Voltage is determined by the amount of Current flowing through the Cathode Resistor. Removing a tube, reduces the amount of current flow, and changes the Bias Voltage for the worse.

CLASS-A CALCULATIONS

To compute the power output of a Class-A Push-Pull pair of tubes, find the point on the Plate Voltage axis corresponding to 60% of the operating Plate Voltage. For a 5D3 Tweed Deluxe with 6V6's, this would be 60% of 370 volts, or 222 volts. Looking at the Characteristic Curves for a 6V6 in your tube manual (you do have a

tube manual, don't you?), go up to the Zero Bias line, and then over to the Plate Current line at 115 mA. This is the Imax value used for the following formulas to determine the Power Output (Po) for two tubes in Push-Pull, and the Plate-to-Plate Resistance (Rpp) needed for the Output Transformer.

$$Po = (Imax \times Eo) / 5$$
$$8.51 \text{ watts} = (.115 \times 370) / 5$$

I measure 9.5 watts out of this amp, so we're close with our formula, and:

$$Rpp = 4 \times (Eo - (0.6 \times Eo)) / Imax$$
$$5,148 \ \Omega = 4 \times (370 - (.6 \times 370))/.115$$

Let's work backwards and take the maximum Plate Dissipation of 12 watts divided by 370 volts and find 32.4 mA to be the maximum amount of Plate Current we can allow.

$$12 \text{ watts} / 370 \text{ volts} = .0324 \text{ amps}$$

Now, find the Bias Voltage line needed to draw 32.4 milliamps at 370 volts. It's -16 volts. To produce -16 volts at .0648 amps (.0324 x 2, for 2 tubes) you need:

$$16 \text{ volts} / .0648 \text{ Amps} = 247 \ \Omega$$

This is a starting point. The 5E3 Deluxe actually uses a 250 ohm Cathode Resistor, and has 19 volts on its Cathodes, so again, we're close.

$$19 \text{ volts} / 250 \ 2\Omega = .076 \text{ A}$$
$$.076 \text{ A} / 2 \text{ tubes} = .038 \text{ A per tube}$$
$$.038 \text{ A} \times (370 - 19 \text{ volts}) = 13.3 \text{ watts}$$

This brings us awfully close to the math, but the best way to work this out is to have a bench with a fixed power supply for the 6.3 volt filaments, the 5 volts for the rectifier tube (if needed), and 300 volts for the preamp tube Plates. Then, have variable supplies for the power tube Plate Voltage, and power tube Bias Voltage. Remember, bias the tubes for their maximum dissipation at idle. When done properly, the Current through the tubes should remain almost totally constant.

Remember, the Plate Voltage, in a Cathode Biased amp, is equal to the Plate Voltage, minus the Cathode Voltage. Multiply this by the Plate Current, and you'll find the Plate Dissipation. In this case, it's 13.3 watts.

A CLASS-A TWIN
(A Really Bad Idea!)

I've modified a Twin to operate in Class-A. This was done by an empirical process known as successive approximation. Using separate power supplies, the Sovtek 5881 tubes were warmed up for 60 seconds at full filament voltage. Then the Plate Voltage was slowly increased from zero, while monitoring the current flow through a single tube by attaching a 1 ohm resistor between the Cathode (pin 8), and ground, and reading the DC voltage drop across it with a millivolt meter.

The goal was to raise the Plate Voltage to 470 volts, without exceeding the 5881's Plate Dissipation rating of 24 watts. This was done by connecting the Positive lead of my Hewlett-Packard DC supply to the amp's Chassis Ground, and the Negative lead to the junction of the two 220K Bias resistors that were disconnected from the amp's negative bias supply. The HP supply allows me to adjust the Bias Voltage from zero to -80 volts in very small increments, and contains A Current Limiting

function to protect the supply if a short circuit should occur.

The target I was looking for was 51 milliamps of current through each tube, at a Plate Voltage of 470 volts.

24 watts / 470 volts = .051 amps

What I was looking for was the Bias Voltage needed to produce 51 milliamps of current flow through each tube, at idle. I found it to be -44.8 volts. You can use the tube's Characteristic Curves to calculate this, except the voltages on the curves are lower than the ones encountered in guitar amps, and if you used a nomograph to convert these, they are so far apart, the results are inaccurate.

Now, I know the Fixed Bias Voltage needed to operate the 5881's into Class-A. The next step is to subtract this Bias Voltage from the Plate Voltage:

470 volts - 44.8 volts = 425.2 volts

Now, I'm going convert the biasing scheme from Fixed Bias to Cathode Bias. When I do this, the Cathode will become 48 volts positive. The effective Plate Voltage equals the Plate Voltage minus the Cathode Voltage. So, the effective Plate Voltage is now 425.2 volts. Reset the Plate Voltage supply to 425.2 volts, and reset the Bias Voltage to draw 56.4 milliamps.

24 watts / 425.2 volts = .0564 amps

The new Bias Voltage needed to draw 56.4 milliamps through a 5881 with a Plate Voltage of 425.2 volts is -38.6 volts. Successive approximation would say to take the 470 volts, minus the 36.6 volts, and repeat the process, but we're close enough. Now to compute the Bias resistor:

.0564 amps x 4 tubes = .226 amps
38.6 volts / .226 amps = 170 ohms

To convert the amp from Fixed Bias, to Cathode Bias, disconnect the Bias Resistors from the Bias supply and attach them to ground. Disconnect to power tube Cathodes (pin 8) from ground, and connect all four of them to a 170 ohm resistor to ground. You can leave individual 1 ohm resistors in place, to make it easier to match tubes.

The Cathode Resistor will dissipate:

38.6 volts x .226 amps = 8.7236 watts

I use a 50-watt aluminum housing resistor, attached to the chassis with thermal grease, and #6-32 screws. I also like to use a rather small 50 uF cap to bypass this resistor to ground. Most amps use a larger cap, but I think the smaller value gives a more lively "bounce" to the sound, and a "wetter" tone. The purpose of this cap is to account for Push-Pull mismatch. In a perfect amp, this cap isn't needed.

Now, slowly fire it up (with a Variac), and double check everything. Definitely install a fan. This amp puts out 60 watts of clean, lush power. Not as hard as you thought?

The Plate Dissipations for some popular power tubes are:

EL84	12 watts	5881	24 watts
EL34	25 watts	6L6GC	30 watts
6V6	12 watts	6550	42 watts

PROBLEMS

(The Plate Voltage is Too High!)

Now, for the surprises. After converting my first Twin to Class-A, I noticed a hum problem that slowly got better as the amp warmed up. The source of this hum was the Output Transformer. The Fender transformers are not built for Class-A use, and after all, they were never meant to. If the windings are not balanced, uneven current will flow through the transformer, and the hum cancellation advantages of Push-Pull are reduced. This is aggravated by the fact that in Class-A, higher current is always flowing through the windings, and any imbalance induces an output signal, even when there is no input signal. In a Class-AB design, the idle current, and hum produced by it, is lower. More hum is produced when a signal is run through it, but that hum is masked by the output signal.

The solution for this was to install a Thunderfunk Output Transformer. The Thunderfunk transformer is designed for 100-watts of output power, at 100% duty cycle. Duty cycle is a measure of how much of the time, 100-watts is flowing through the transformer. The extra heat developed by the full dissipation at idle aspect of Class-A design will probably burn out the stock Fender transformer. Again, it was never designed to do this, and this is not meant as a criticism of Fender.

There might also be a problem with the Fender Power Transformer, as it's delivering full power, all the time. Without knowing what the Transformer's specifications are, I can't tell you for sure. There is a way to measure the winding's resistance cold; run the amp, and measure the winding's resistance hot.

Temp Rise = (hot R - cold R) / cold R x .004

The temperature coefficient of copper is 0.4% per degree Celsius. The change in resistance can be used to calculate the transformer's internal temperature rise, which will tell you if you're overheating the transformer. Of course, we would still need to now what temperature the transformer was designed to operate at, to know if we were stressing it.

Since, I used a Thunderfunk transformer, your measurements won't match mine exactly. The Thunderfunk transformer has a 4K5 primary impedance which reduces the current flow, and makes it easier to meet the Class-A dissipation target. Using these formulas though, you'll be able to design your own Class-A amp.

CLASS-AB POWER AMPS

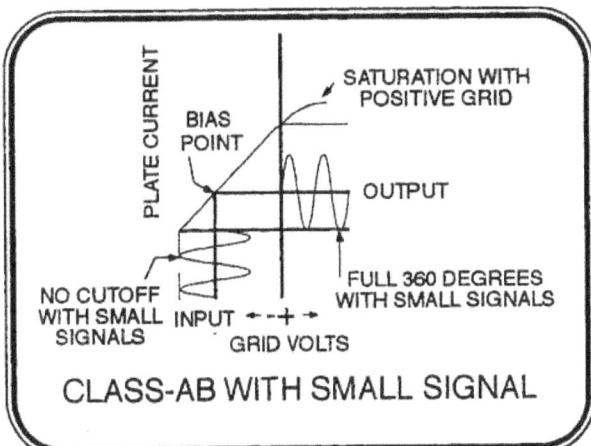

CLASS-AB WITH SMALL SIGNAL

With Class-AB design, the Plate and Screen-Grid Voltages can be made higher than for Class-A designs because the increased Negative Bias holds Plate Current within the tube's Plate Dissipation rating. The use of higher Plate Voltages allows higher output from these designs,

and that's why they put out more power than Class-A designs. Remember, Current is power, Voltage is potential power. If you can raise the Plate Voltage you can get more power out of the tube, and the Negative Bias keeps the Current from burning it up.

I know if you think about Ohm's Law, it doesn't make sense. If the Voltage goes up, the Current must go down to balance the Maximum Dissipation Rating. They're both multipliers, and increasing the Voltage shouldn't result in more Power from the tube, because we're still talking about Maximum Dissipation. Don't confuse Maximum Power Dissipation with Maximum Signal Voltage. What we're saying is the tube can't throw the Voltage due to a low Supply Voltage. Raising the Voltage allows the tube to throw a larger output, generating Plate Dissipation closer to the Maximum. You see, we have Output Clipping, AND Input Clipping. The higher Grid Bias in Class-AB, also allows a higher Input Signal Voltage to be used without Grid Current being drawn, and for both reasons higher power output can be obtained compared to Class-A service.

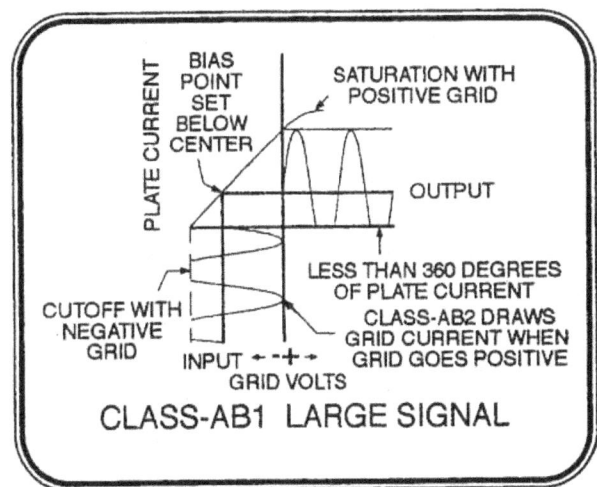

CLASS-AB1 LARGE SIGNAL

In Class-AB design the Power Supply needs to have better regulation than needed with Class-A design, for reason explained earlier relating to sag. Otherwise, the fluctuations in Plate Current that occur with Class-AB operation, could cause fluctuations in the voltage output of the power supply, resulting in decreased power, increased distortion. As should be obvious, both Class-AB and Class-B designs require a balanced output stage using two tubes.

CLASS-B POWER AMPLIFIERS

A Class-B amplifier employs two tubes connected in Push-Pull, biased so that the Plate Current is almost zero when no signal voltage is applied to the grids. Because of this low value of no-signal Plate Current, Class-B amplification has the same advantage as class AB, that is, large Power Output can be obtained without excessive Plate Dissipation.

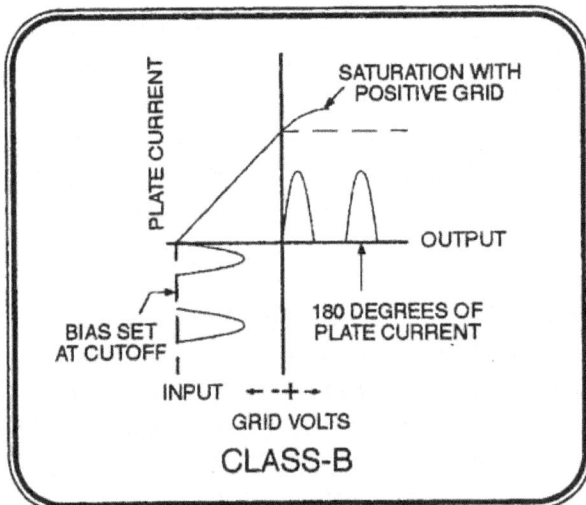

CLASS-B

Class-B operation differs from Class-AB in that Plate Current is cut off for a larger portion of the negative Grid swing, and the signal swing is usually larger. Because tubes designed for use as Class-B amplifiers usually operate at zero or low bias, each Grid is at a positive potential during all or most of the positive half-cycle of its signal swing and consequently draws considerable Grid Current. There is, therefore, a loss of power in the Grid circuit. This condition imposes the same requirements on the driver stage that a Class-AB design does. The driver needs to deliver considerably more power than required for the Class-A grid circuit in order to develop full power and keep distortion low. Musicman amps run Class-B. That's the reason they can get away with 700 volts on the Plate, and can get 112 watts from four 6L6's.

There are several tube types designed especially for Class-B service. The characteristic common to all of these types is a high Amplification Factor. With a high Amplification Factor, Plate Current is small even when the Grid Bias is zero. Therefore, these tubes can be operated in Class-B at a Bias of zero volts, so that no bias supply is needed. A number of Class-B amplifier tubes consist of two triodes. The triodes can be hooked up in push-pull so that only one tube is required for a Class-B output stage. An example of a Class-B Twin Triode is the 6N7.

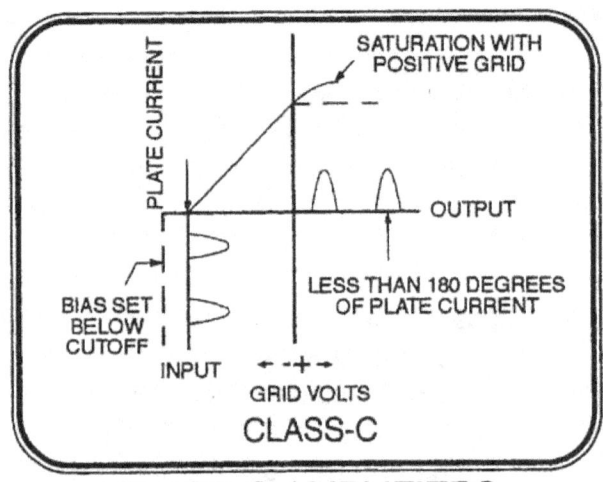

CLASS-C

CLASS-C AMPLIFIERS
Class-C biasing is not used in audio.

CHAPTER 12
THE ORIGINS OF AMPLIFIER TONE

The first stage in a guitar amp is a Preamp, and is the most important stage in the amp. This is where the signal from the pickup is captured, and raised to "Line Level."

Line Level originated with the telephone company as a standard audio signal level for telephone "lines." Line level is notated as 0dBm. The "m" in dBm stands for milliwatt, and reminds us that we're talking about 1 milliwatt (.001 watts) of power, dissipated into a 600 ohm telephone line. 0 dBm equals .775 Volts.

.775V / 600 Ω = .0012917 A of Current
.0012917A x .775V = .001W or
1 milliwatt dissipated into 600 ohms

A 0dBv level equals 1 volt, and is also used as a nominal operating level.

A Power Amp expects to see .775 Volts, or approximately 1 Volt, "Line Level" in. If your signal source can't supply 1 Volt, you'll need a "preamp," or an amplifier "pre" (before) the amplifier to bring the Voltage up to "Line" level. A Preamp is a Voltage amp, and amplifies signal voltages. A Power Amp is a Current amp, and is designed to convert signal voltages into an output Current sufficient to drive the speakers. Stages in between are called Drivers, such as, Reverb Driver, Tone Driver, etc. Sometimes the Gain Stage follows the loss of Gain and then are called Recovery Amps, such as Tone Recovery. Gain had to be added to the early amps before they could have things like Tone Controls. The Tone Controls

actually throw Gain away, and need to be compensated for.

Now, you understand that a Preamp is just an amp that is designed to operate with a small input signal voltage. Some phono cartridges (such as moving coil) put out extremely small signals, and need a "headamp" before the "preamp." Since the average guitar pickup puts out about 150 millivolts, you'll need a "preamplifier" to raise the 150 millivolts to the 1 volt "amplifier" level. The hottest pickup I've ever measured was in a stock 1959 ES-335, and it put out .800 Volts. If run wide open all the time, it could drive the input of a Power Amp directly.

If you plug a CD player into the input jack of your guitar amp, you'll most likely get distortion. The CD player puts out Line Level already, and you're plugging it into another Preamp. If you think about it, a CD runs off of a Digital to Analog Converter circuit. Already being a circuit, it has Gain built into it. Preamps are for Sensors, such as guitar pickups, or phonograph needles. The Preamp in a CD player, amplifies the signal off the LED pickup inside the unit, and is already self-contained.

Most guitar amps will only take about .350 volts in before the first stage overloads. If you play Steel Guitar, you'll know what I mean. Without a volume pedal to back these babies off, they'll make most amps puke. Red Rhodes wound a pickup on Clarence White's Tele, the one that's owned, and used by Marty Stuart.

Marty said the pickup is extremely hot. I suppose being a Steel player, Red wound it that way. The Triggs amp I build will take .850 Volts in before clipping. This helps maintain a clean, dynamic quality to the amp. As I said, the input stage is the most important part of the amp for capturing the detail of the pickup.

CLEAN TONE

Have you ever heard someone say, "Should I buy new speakers to improve my tone?" Can a good speaker recover tone lost by the amp? If the first stage doesn't work right, you'll never get good tone downstream. Of twenty-eight 100-watt amps recently reviewed, the most common comment seemed to be, "There was no clean tone," and "The tone controls didn't work." To save face on one of the amps, the staff said, "(The) tone control knobs don't provide extreme tonal versatility, but no matter where they're set the amp sounds good." What is being ignored here is the amp was being tested in a small, carpeted office. What happens when you NEED tone controls, like on a real world job? And sounding good no matter how the ineffective tone controls are set, means, the amp only has one sound, and either you like it, or you don't. "Like any other amp with such distinct character, you're either going to love or hate (it)." Hmmm.

Many amps today ignore basic versatility in order to accomplish one, possibly good, or bad, distortion tone. Being a self-fulfilling prophecy, a common belief is, "You can't get more than one good sound out of any amp." I believe that you can't get a good dirty tone, if you don't first have a good clean tone.

MECHANICAL vs. ELECTRICAL vs. COST

You can make electronic parts cheaply, but you can't make good mechanical parts cheaply. This should be obvious when you consider we run cars off computers, because a mechanical carburetor that could make all of the adjustments necessary to meet modern pollution control requirements would be very expensive, or even impossible to build.

You might also consider that a transistor sells for a few cents, while a vacuum tube sells for a few dollars. Transistors are made by a process similar to silk-screening, while tubes are electro-mechanical devices, very much dependent on their mechanical construction. When you wonder why a tube amp cost more than a solid state amp, this is one reason. Another reason is the cost of the transformers. A solid state amp doesn't need the output transformer.

WEIGHT vs. COST

When looking at VCR's, pick them up. The heavier one is the better one. It has a stronger, and truer mechanical system in it. Sometimes the only difference between CD players is the quality of their mechanical systems. The one with better bearings, or a heavier alignment system will have better performance, and last longer. In "Jurassic Park," during a scene in the Ford Explorer, the lawyer asks about the night vision glasses the boy found. "Are they heavy? Then they're expensive, put 'em back." I guess I'm not the only one making this point.

When considering the cost of an amp, consider its weight. It makes sense that the heavier amp, at the very least, has more raw materials in it. If a tube amp weighs twice as much as a solid state amp, and cost twice as much, doesn't it make sense that it's because it has twice the amount of steel, and copper in it? When you get into very heavy amps, you'll find aluminum. Aluminum costs more than steel, but helps reduce the total weight of the amp.

DESIGN PRIORITIES

The first part of a good amp is the actual circuit design. I heard Robert Plant on Centerstage (PBS) say that when he was with Led Zeppelin, he could never hear the band, because their equipment back then was so bad. The amplifier circuits used back then were from the 1930's. They evolved from systems made for the talking movies. Almost every guitar amp made today is a copy of these very old designs. Thunderfunk amps are based on designs from the 1940's. They are extremely modern by comparison.

This gives my amps something that nobody else has, cohesiveness. Back up 50 feet from the stage, and try and listen to what the keyboard, vocals, drums, and bass guitar are doing. Now listen to the guy with the Thunderfunk amp. He'll be bouncing balls of tone off the back wall. Cohesiveness is the ability of the amp to project your notes without the blur associated with the other instruments. The effect is you can cut through without turning up your volume. This is readily apparent even on my demo tape. The guitar rises out of the mix.

Next in importance is the tonal equalization of the design (the EQ), determined by the value of the parts. By changing one cap value, you can improve the sound of an amp, or destroy it.

Next, is the quality of the parts. Ceramic disk capacitors change value with frequency, making them poor for use in tone circuits, but they're cheap. I can hear when ceramic caps are used in an amp. They have a bright, brittle tone.

In addition to capacitance, capacitors also have resistive, and inductive losses. High quality parts can keep these losses at a minimum. If you see ceramic caps in your amp's tone circuits, you can bet the manufacturer was more interested in profit, than tone.

TOLERANCE

Least important is the tolerance of the parts. If you have Inverter Coupling Caps that are closely matched, but are the wrong value, their tight tolerance won't mean a thing. And what about the tolerance of your transformer? My transformer company tells me that a perfectly built transformer has a tolerance of 25%. And I've seen inverter tubes with two sections that were out of wack, and it sounded like a flat tire, or as bad as a blown speaker.

MATCHING

George Alessandro, of Hound Dog Amps, pointed out that the difference in balanced Inverter cap values is a Frequency Matching, NOT a Power Matching. The difference in Cap value, affects what Frequency the turnover point is. Once you get into the higher frequencies, the impedance of the cap is so low, any mis-

the guitar's frequencies, the slight difference in cap value, will not be noticeable. I'm amazed (not really) when I hear claims made, "Matched Inverter Coupling Caps really improved the quality of the amp!" Maybe in your dreams.

I agree with Bill Holter, of Vintage Sound. He thinks the oldest Marshalls sound the best, BECAUSE of the poor tolerance of the parts. I can understand this, because if you get an amp out of balance, it generates one distortion tone on the top tubes, and a different response on the bottom tubes, with the two tones blending together to form a third tone. This adds richness to the distortion blend. A term for this is synergy. One plus one equals three.

Steve McClure (Garth Brooks' Steel & Slide player) blew two 30-watt Celestions in his 4x12" cabinet. We replaced them with two 100-watt Celestions. The idea behind this is the 30-watters don't have much of a low end, and they break up easily. The 100-watters have a low end, and take more push before breaking up. The blend of the two speakers provides a cabinet with a low end, that breaks up a little bit as the 30-watters are pushed. As you push harder, the 100's start to break up. This gives you a wider range of distortion response. A wider range of touch sensitivity. Remember, playing on the edge? Same thing. Another example of an advantage to being unmatched, and unbalanced.

DISTORTION TONE

Another point I'd like to make is, I think it's exaggerated that power amp distortion sounds better than preamp distortion. You can find Marshall distortion in their pre-

amps. And if you don't like preamp distortion, what about Clapton's Soldano tone? I would think that preamp Triodes, operating in Class-A, would make better distortion amps than power Pentodes operating in Class-AB Push-Pull, which are designed to be more linear than Triodes. The only thing missing from the preamp circuit is the interaction of the Output Transformer, and the Voice Coil. To understand this, let's talk about Reflected Impedance.

REFLECTED IMPEDANCE

My family has an expression that says, "It's technical, YOU'D never understand it." Some service guys use Reflected Impedance as a way of saying, "It's Technical, so don't ask me to explain it." Let's have a laugh on them by explaining it to you.

It's really quite simple. A Current flowing through a Coil creates a Magnet. A Magnet moving through a Coil creates a Current. When a signal flows through the Primary of the Output Transformer, it induces a Current into the Secondary (Speaker side) of the Transformer. The Current flowing through the Secondary, induces a Signal back into the Primary. After all, Transformers work both ways.

When a signal flows through the Voice Coil, it causes the Voice Coil to move through the Speaker's Magnet. The Coil moving through the Magnet is the same as moving the Magnet through the Coil. It induces a signal back into the Coil. This signal flows back through the Transformer Secondary, inducing a signal back into the Transformer Primary, and you have what some people call Reflected Impedance.

It's just like using a speaker as a microphone. If you talk into the speaker, you'll produce a signal at the voice coil. It's like those sci-fi movies were the TV's watching you. When you talk in a room with piped in music, everything you say is picked up by the speakers. They could be listening to you. Now that you know this, we'll have to kill you.

Reflected Impedance is really a term used to explain that a transformer has no Impedance of its own. What the Primary Circuit sees is the Secondary Impedance "Reflected" through the Transformer. Of course, that Impedance is "Transformed" (converted) to a different value by the operation of the transformer. So, the Primary sees 8 Ohms reflected through the transformer and transformed to a higher value by the winding ratios.

STUDIO SECRET

To get a big sound, use a small speaker. Brad Davis (Marty Stuart's Rock & Roll Cowboys) built himself a small open-backed box, with a 8" speaker. He drives it with a 100-watt Thunderfunk amp. It records great. I think there are two reasons for this. The first is cone movement. A 12" speaker, at low volume, has a very shallow cone movement (excursion). You're asking the microphone to be sensitive to this delicate movement. At the same volume, a smaller speaker has a larger cone movement, causing it to "pump" the air better, making it easier for the microphone diaphragm pick it up.

The second reason is the larger cone movement causes more signal to be generated by the voice coil's movement, which kicks more signal back into the Output Transformer, affecting the tone of the amp.

CONCLUSION

Ultimately, the tone is in your fingers, or in the case of a steel player, in his slide (just kidding). Now that you've seen some of the design decisions that go into an amp, as they relate to tone, let's examine the actual amplifier circuits, to learn what each stage actually does.

CHAPTER 13
THE PREAMP

The individual parts of an amplifier circuit have specific names that allow you to understand the purpose of the circuit, no matter what amplifier you're working on. Let's start at the beginning and go through all the amp's circuits.

THE INPUT CIRCUIT

Most amps have two input jacks per channel. Each one is connected to a 68K resistor, with a 1M resistor (Grid Leak Resistor) connected to ground. If you plug into Input Jack number one, the 1M to ground comes first, with the 68K connecting to the Grid of the first preamp tube. If the 68K came first, it would act as a divider, the equivalent of a 1,068K pot turned down to the 1M point.

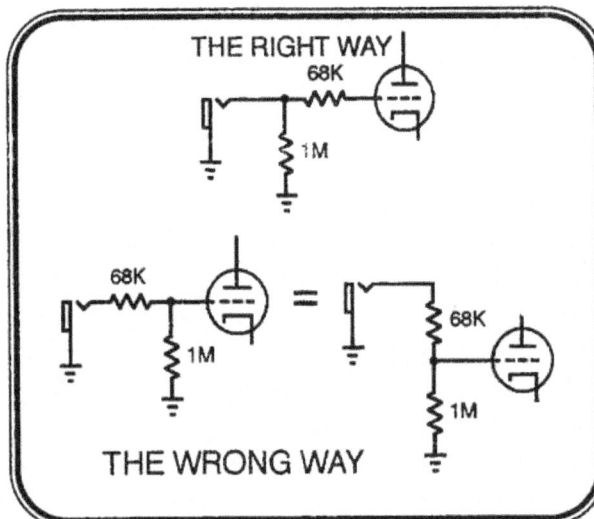

THE RIGHT WAY

THE WRONG WAY

If you plug into Input Jack number 2, the signal goes through the first 68K resistor to the tube's Grid. With nothing plugged into Input Jack number 1, its 1M resistor is shorted out, and its 68K resistor is con-

nected to ground. This is the equivalent of a 136K pot turned down to the 68K point, or halfway off.

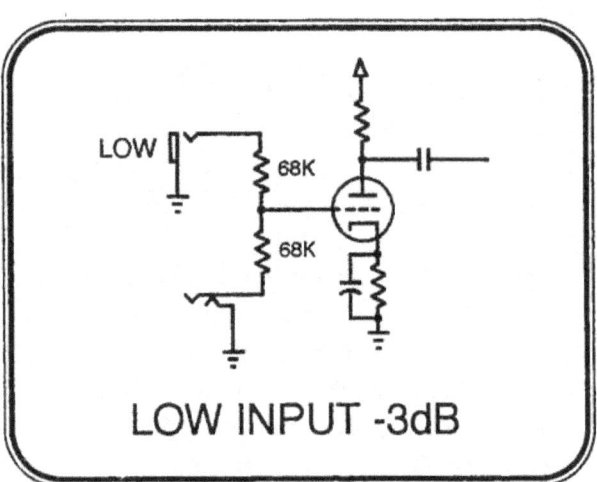

LOW INPUT -3dB

This provides a loss of half the input signal, or a 3dB loss. That's why the number 2 input is quieter than the number 1 input. Half of its signal has been thrown away. If you run active pickups, this can help you avoid clipping the input stage. If you plug in two guitars, the 1M resistor is back in the circuit, and both inputs revert back to full gain.

ISOLATION RESISTORS

On older Tweed amps, there were no 68K Input Resistors. On some Tweeds, there's also an Input Capacitor to block DC from draining away through the guitar's pickup coils, and is needed there for the Grid-Leak Biasing scheme used onsome of these early amp. The reason there's no Input Resistors on these designs is due to the use of a separate Triode tube for each input. The 68K Resistors are actually

94

used to "Isolate" one guitar's pickups, from the other. If a pickup was directly connected to the tube's Grid, and another pickup was also connected, the two pick-ups would "dump" signal into each other's coils. If most pickups are between 5K and 8K, then 6K8 would be an intermediate value. Remembering our rule of 10's, 6K8 becomes 68K. This allows each pickup to drop it's signal across a 68K resistor, putting 68K x 2 between the pickups.

If each input is connected to its own grid, as on some Tweed amps, the channels are mixed downstream. We'll be running into this Factor of 10, Isolation Resistors, and channel mixing, again, as we move down through the amp circuit.

THE PREAMP

This first stage can be called The Preamp of the total circuit, although all the stages operating at, or below Line Level, are technically Preamps. Since Preamp stages further downstream, perform specific functions that carry titles, like Tone Recovery Amp, we'll just call the first stage "The Preamp."

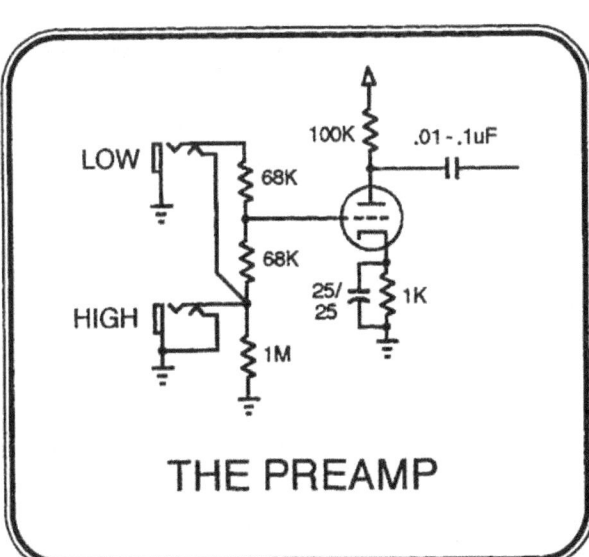

THE PREAMP

The Preamp raises the small signal from the guitar's pickups, typically 150 millivolts (or .150 volts) to about 4 volts for a 12AX7. Because audio signals oscillate, like sine waves, the voltage is called an AC Voltage (Alternating Current Voltage). I use a Fluke 85 meter that measures the voltage 10 times a second, and records it as a maximum, minimum, and average voltage. By plugging this into a guitar, and banging on it as hard as I can, I can record the pickup's maximum RMS output voltage.

When I'm testing an amplifier, I set the signal generator (a piece of test equipment that generates a continuous, and steady signal, of sine, square, or triangular waves, at a selectable frequency, and selectable AC voltage level) to 1,000 Hertz (cycles per second), sine waves, at a voltage level of 100 millivolts (.1 volts). This signal is amplified by the first stage preamp, and you should see an AC Output Voltage, at the tube's Plate, of 4 to 6 Volts, AC.

In a Fender amp, the input goes to pin 2, and the output is pin 1. Be careful at pin 1. It has 150 to 300 Volts DC on it, as well as the 4-6 Volts AC. If you find less than 4 AC Volts at pin 1, the tube is weak, and should be replaced. If the tube is a 12AT7, or a 12AY7 (as in tweed amps), the normal output voltage will be more like 2 AC Volts.

If you want to check the gain on the other triode in the tube (9-pin miniature preamp tubes have two triodes in them), adjust the Volume control so that you have 100 mV on pin 7 of the tube. Measure the AC Voltage at pin 6 (also a High Voltage Pin, Be Careful!), and look for 4-6 AC Volts.

THE PLATE RESISTOR

The Plate Resistor of a preamp stage is traditionally 100K. If you make it larger it will slightly increase the gain. The range of values is 47K to 330K. To fool with it, use a pot (remember, the pot's going to have high DC voltage on it, be careful), and sweep the range. My advice, leave it at 100K.

THE GRID LEAK RESISTOR

The Grid Leak Resistor is typically 1M to 5M. Its purpose is to allow electrons stuck on the Grid to "Leak" away to ground. (Remember, there's no DC Grid Ground connection inside the tube). Sometimes the Grid Leak resistor is part of a Voltage Divider from the previous stage. In these cases, it might be as low as 10K. It doesn't really matter how low it is, except, the smaller the value, the more Input Signal is lost to ground.

THE CATHODE RESISTOR

The Cathode Resistor is the main gain setting component of a tube preamp stage. It can range from 820 ohms to 100K. Below 820 ohms, the tube will not sustain bias, and the gain will drop off. Above 47K, gain is basically Unity (1 volt in, 1 volt out). Try replacing the Cathode Resistor with a 50K pot, and adjust to hear the difference. If you run a test signal through the amp, watch the output of the Triode to see the effect of the change in value. You'll notice one side of the wave clipping before the other. This is a minimum setting. Above and below this, the gain drops off. I personally believe 3K3 is the best SOUNDING value for Cathode Resistors, although I like to use 1K for the first stage. This is based on low noise and maximum input signal level before clipping the input.

THE CATHODE CAPACITOR

You can double the gain of the triode stage by adding a Cathode Capacitor across (parallel to) the Cathode Resistor (review "Decoupling Capacitors," in the chapter on capacitors).

Basically, as the Input Signal increases, making the Grid more Positive, the tube starts to conduct more current. The current flows through the Cathode Resistor causing and "IR" Drop, which makes the Cathode go More Positive. This Positive Cathode Voltage makes the Grid relatively Negative, setting up the tube's Bias (Self-Bias, or Cathode Bias).

The problem is, this Bias Voltage varies with Current flow through the Cathode Resistor. A small signal has low Cathode Current flow, and therefore a low Bias Voltage. A large signal has large Cathode Current flow and a high Bias Voltage. This means the Bias Voltage varies in sync with the Input Signal. This AC Voltage is the equivalent of Ripple on the Bias Supply. The AC Voltage on the Cathode counteracts the signal by tracking with it. This is called "Degenerative Negative Current Feedback." The purpose of the Cathode Cap is to increase Gain by decreasing this Negative Feedback. You can think of it operating in two different ways.

As you know, Caps have relatively low AC Impedance, and theoretically infinite DC Impedance. The first way to think of the Cathode Cap is as a Low AC Impedance path to Ground. It eliminates the AC Cathode Voltage, by giving it a path to Ground. The DC Voltage is blocked by the Cap, and only flows through the Cathode Resistor.

The second way to think of the Cathode Cap is as a sort of short term storage device. As the Signal causes more Current flow, and a higher DC Cathode Voltage, some of that Voltage is stored on the "top" plate of the Cathode Cap. During periods of low Input Signal, when the Cathode Voltage is dropping, this "stored" Voltage is returned to the Cathode, helping to maintain a higher average DC Cathode Voltage.

There are some special types of capacitors that store Voltage for so long, that they act as Batteries to backup CMOS (Complementary Metal Oxide Semiconductor Field Effect Transistors (FETs)) chips in computer memory circuits. A Voltage is stored on the Cap when the computer is on, and then the Voltage is maintained in the Cap when the power is turned off, continuing to power to the memory circuits. Part of the reason this works is the Very High, Power Supply Impedance (Low Current requirements) of CMOS type memory. My point is, thinking of a Cap as a Battery is a valid way to conceptualize it.

FREQUENCY RESPONSE

Here's the cool part. If the Cathode Cap is made small, it loses its decoupling properties at low frequencies. It's just not big enough to store Voltage long enough, or provide a low enough Impedance path to Ground at low frequencies. If the Cathode Cap "disappears" at low frequen-

cies, the low frequency gain will be reduced. I made a point of relativity earlier in this series. Here it comes again. Reducing the low frequency gain is the same as increasing the high frequency gain. Marshall amps use Cathode Caps of .68μF. Fender amps use 25μF. The small Cathode Cap in Marshalls, strips out the low end from the first stage. This is important in a distortion amp, as the low end will cause a "roll" and add a muddiness to the distortion.

HOW TO MAKE AN AMP SOUND BETTER

Toby Seay, an engineer in Nashville, told me his three rules to make an amp sound better.

1. Make It Brighter
2. Make It Louder
3. Make It Louder & Brighter

This "trick" is used by some amp companies to fool the casual listener into thinking they make cool amps. You can easily mod a Fender amp by changing the 25uF caps to .68uF. This will definitely make the amp brighter. It will also destroy the low end, and give the impression of a lack of power. The reason the Cathode Cap trick works with Marshalls, is they have two gain stages before the Tone controls. Fender only has one. It may seem trivial, but one of the "secrets" of the Marshall sound is the Volume control determines how much gain is pumped through the Tone Controls. While in a Fender, the Volume control determines how much of the output of the Tone Controls is passed on to the Tone Recovery amp. This is covered in the next chapter.

It's like this. In a Marshall, you have a Gain stage; a Volume control; another Gain stage; and then the Tone Controls. In a Fender, you have a Gain stage; Tone Controls; Volume Control; and the Tone Recovery amp.

The advantage of Fender having two gain stages is lost by the amount of gain lost in the Tone Controls, and the Volume Control, before you even reach the second stage Tone Recovery amp.

With a Marshall on "10" it's as if there was no Volume Control. This results in a gain stage overdriving another gain stage, and then this distorted signal is tuned by the Tone Controls. Of course, the Marshall tone controls don't work very well, because so much tone was stripped out by the earlier gain stages, as a means of controlling the distortion. This leaves very little tone for the tone controls to work with.

With the advent of distortion playing, you can see why Marshall kicked Fender's butt. Compare a Marshall 100-watt head with a Showman. The Fender loses half its gain in the Vibrato circuit. This is why disconnecting the Vibrato is a popular mod. Then throw away more gain in the Tone Controls. Then throw away more

gain in the Volume control, before you finally hit the second gain stage. Turn it up to eight, if you want it to distort. If you think this is unfair because a Super Reverb distorts at 3 or 4, remember, it's a thicker, throatier distortion, partly caused by the 25uF caps, and partly caused by it being a 40-watt amp. The 10" speakers help to brighten it up. What do you think sounds louder? A 50-watt Marshall head (34-watts actual watts), or a Super Reverb? In defense of the Fenders, Marshall is not known for its clean sound. Of course, this is an over-simplification of a complicated subject. And now back to our program, already in progress.

COUPLING CAPS

Another way to thin out the distortion tone is to use a smaller coupling cap. Change a .1μF to a .001, and you hear a significant reduction in bass response. This is exactly what the FAC switch in Orange amps does. It changes one of the coupling caps from .068μF down to 330pF in six discrete steps. At every step, the tone gets thinner.

Next, let's look at Gain Stages, and try to answer that endless question, "How Many?"

CHAPTER 14
GAIN & TUBE DISTORTION

MARSHALL v. FENDER

What sounds louder? A 50-watt Marshall (34-watts actual watts), or a Super Reverb (40 actual watts)? In defense of the Fenders, Marshalls are not known for their clean sound. Of course, this is an over-simplification of a complicated subject. Let's start with the simplest amps, the single gain stage designs.

SINGLE GAIN STAGE AMPS

The simplest amps are the single gain stage amps, like the non-Top Boost Vox AC-30 and AC-15 amps. I suppose Vox invented Power Amp Distortion. With a single Gain Stage, Volume Control, Inverter, and Power Tubes, it's the essence of simple design. These simple designs are all over Europe. Most Americans don't realize that in Europe, conservation of natural resources is a prime design factor. When you look at a European sink, you'll marvel at its "sculpted" design. American sinks look like blocky blobs of porcelain. European sculpting reduces the amount of porcelain needed to form the sink, preserving natural resources. Even if it's harder to form, the Europeans will do it.

The savings in material has a double effect; less material is used in the first place, and then in Europe, things are not thrown out, and replaced, just because they're old. When you're staying in a 200 year old hotel, the 50 year old elevator isn't really a problem. To make this hap-

pen, maintenance becomes a big deal. As with Vox, you might wonder why English cars are finicky in their adjustment. I think it's because the English like to fiddle with things. Real Vox amps (not reissues) are finicky and difficult to service properly. I would recommend caution in selecting a service shop.

TWO GAIN STAGES

The normal preamp gain structure of a Fender, Marshall, or Vox amp is two gain stages. The difference between Fender and the English designs, is the English use a first gain stage feeding a volume control, which then drives a second tube. Following are the Tone Controls, with no additional gain into the Inverter/Driver.

This design allows the player to turn the volume up to the point where the second gain stage starts to distort. This is all but impossible with a Fender, because the Tone Controls AND the Volume Control throw away so much gain before the second gain stage is met, it's nearly impossible to get the Blackface and later Fender amps into high gear.

THREE GAIN STAGES

But if you add another Gain Stage to a Fender after the Tone Recovery amp, plenty of drive can be provided into the Inverter/Driver tube. With this one Gain Stage added after the Tone Controls (Post-EQ), you have a 1+2 type amp. The Blackface Bassman is a 1+2 type amp, showing why it's a popular amp to modify, already having an extra gain stage that can easily be used to turn the amp into a "high gain" model. Boogies are 1+2 designs as is the Trainwreck Express. I build an amp called the "Nash" that uses the 1+2 layout, and it goes from clean to dirty on one knob.

The Bandmaster amp is harder to modify, without stealing the Vibrato tube, or one of the Channel tubes, for the extra gain stage. The Fender Reverb Amps are easier to modify in this respect, if you steal one side of the Reverb Driver tube for the extra gain stage. Without doing this, you would have to steal the Vibrato tube, or one of the Channel tubes, or punch the chassis and add another tube. The only room on the chassis where you can add another tube is under the fiberboard, since the back row is already taken up with tubes and the reverb transformer. You'll have to relocate some parts and shorten the fiberboard to make room.

THREE GAIN STAGES WITH MASTER VOLUME

Add one Gain Stage in front of the Tone Controls (Pre-EQ), and you have the Marshall Master Volume design. A total of three gain stages, all three in front of the EQ, and none behind. You see that I'm making note of how many Gain Stages, and where they're located in the schematic.

I'm going to divide all designs into two major classes; Pre-EQ, and Post-EQ design. To identify these different topographies, we'll use a notation of 3+0 for the Marshall above. Some different amps would then be:

Vox AC30	1+0
Marshall	2+0
Vox AC30 Top Boost	1+1
Fender	1+1
Nash, Clean	1+1
Trainwreck Rocket	1+1
Blackface Bassman	1+2
Boogie	1+2
Nash, Dirty	1+2
Trainwreck Express	1+2
Marshall MV Low Input	2+0
Marshall MV, Lee Jackson	2+1
Thunderfunk 100ELS, Clean	2+1
Soldano SLO-100, Clean	3+0
Marshall MV, High Input	3+0
Marshall MV, Frank Levi	3+0
Marshall MV, Soldano	4+0
Thunderfunk 100ELS, Dirty	4+1
Soldano SLO-100, Dirty	5+0

FOUR GAIN STAGES

Add two gain stages in front and you have a Thunderfunk, or Soldano Marshall amp for a total of four gain stages (Pre-EQ).

The more gain stages, the smoother, and more compressed the tone. The four gain stage designs have an instant Allman Brothers tone. The three gain stage designs have a rougher, more aggressive texture to them. I was working on an amp for a guy named Eddie, and he thought the three gain stage, Post-EQ design had too much gain. He preferred the Thunderfunk, which is actually, at four gain stages, the higher gain design, but sounds smoother.

FIVE GAIN STAGES

The Soldano amp has Five Gain Stage Pre-EQ, but the FX Loop is part of it. If you remove the FX Gain Stage, your back to four distortion stages.

TUBE DISTORTION

To explain the effects of these gain stage structures, and tube distortion in general, we're going to take a short detour. Hope you stay with me until we come out at the other end.

TUBE EMULATORS

Recently, an article by Brian Santo appeared in "Technology" magazine, on the subject of Tube Emulators. It follows the same line as my chapter from this book called, "Are Tubes Louder Than Transistors?" The basic idea is that tubes and transistors sound the same when run undistorted. When the circuits are clipped, that's when the tubes work their magic. But another factor was raised that I had never heard before, but makes interesting sense. "A tube amp clips with a Duty Cycle that varies as a function of drive level."

DUTY CYCLE

To understand Duty Cycle, think of a Sine Wave with a perfectly matched and symmetrical, top and bottom half-wave. The top half-wave is exactly one half, or 50%, of the total signal, and the bottom half-wave is the other 50%. This is a Duty Cycle of 50%.

If the top half-wave lasts for more, or less than 50% of the duration, it's Duty Cycle is not symmetrical, and it is either more, or less than 50%.

In 1983, John Murphy, of True Image Audio (Escondido, CA), designed a solid

state tube emulator, when he was chief engineer at Carvin. The circuit is contained in Carvin's SX series solid-state guitar amps. Carvin did not patent the circuit, and it is now in the public domain. The way it works is to amplify the signal and use a soft clipper to clip only one side of the wave.

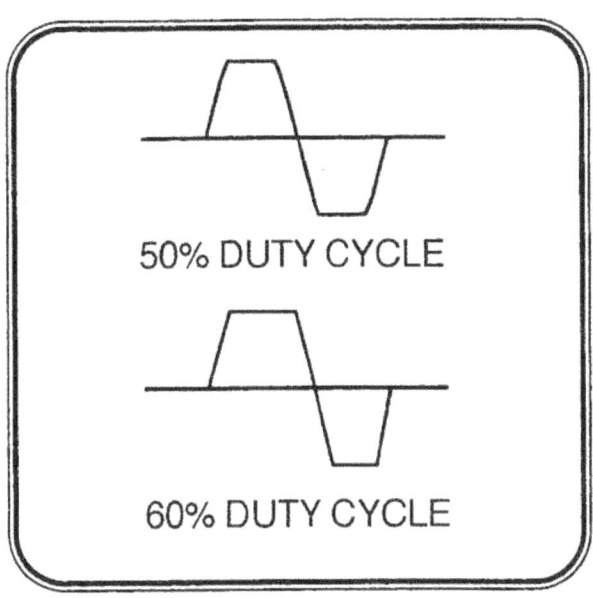

The Asymmetrical Clipper then follows with another gain stage, and a clipper in the other direction. The reason these clippers work, is a Solid State Diode will block Current flow in both directions, until a small amount of Voltage is developed across it. This "Diode Drop" Voltage is usually .6-.7 volts, although it ranges from approximately .2 volts to over 1 volt, depending on the type of diode. If you run an AC signal across two diodes facing in opposite directions, one of the diodes, or the other, will conduct when the voltage reaches .6 volts. This acts to "clamp" the signal, and actually "clips" off the top and bottom of the Sine Wave, producing a Square Wave. The more gain put across the diodes, the more "square" the wave becomes, and the more distorted the output becomes.

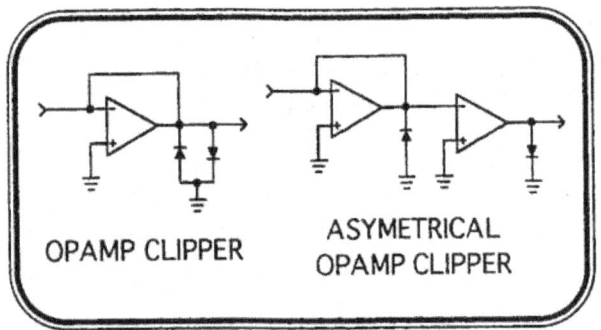

OPAMP CLIPPER   ASYMETRICAL OPAMP CLIPPER

This is a common approach to building signal clippers. There's a modern English amp whose distortion circuit is built around solid state clippers. Also, the Jim Kelley amp used "1N914" diodes in the Preamp to clip the signal.

There's also a company that sells "Tube" equipment, that actually uses an opamp to drive a tube. But, the opamp can't overdrive the tube before the opamp itself clips by hitting its supply rails. So, you end up with solid state distortion "filtered" through a tube, and a claim of "Genuine Tube Distortion."

The Carvin approach is to clip only one side of the wave at a time. Then the signal goes through another gain stage and another clipper, to do the same thing to the other side of the wave. Because there's gain between the two clippers, the signal hits the second clipper at a different level, resulting in an Asymmetrical clip.

This goes back to something I've said about matching and balancing the Inverter circuit. I don't think it's important, and is probably detrimental to the distortion tone. I've thrown amps out of balance on purpose to improve the texture of the distorted signal.

As an example, if a Push-Pull circuit is in perfect balance, it cancels even order har-

monics, and lets a larger percentage of odd order harmonics through. This adds a grittiness to the tone that I consider undesirable.

OPAMPS & POWER SUPPLY CLIPPING

We haven't covered opamps yet, which is a subject in itself, but you need to know a little bit about them now.

Opamp is short for Operational Amplifier. "Operational" refers to Arithmetic Operations like addition, subtraction, multiplication, division, integration, and differentiation. Opamps are actually Analog Computers, as opposed to a Digital Computer.

Opamps can add AC and/or DC Voltages together. An example of this is using an Opamp as a "Summing" Amplifier in a mixer to "Add" all the Channels together to form an Output Buss.

An Opamp can add 1 volt DC and 2 volts DC, and produce an output of 3 volts DC. This is an example of an Analog Computer. The guns on U.S. battleships are aimed with Analog Computers. A dial is turned to indicate the target range, and the position of the dial indicates how much DC voltage is being put into the computer. This voltage is "summed" with other voltages, like ship speed, wind speed, Target direction, wind direction, etc., and an "answer" voltage is created, which is used to position the guns. When the ships were recommissioned in the 1980's, these computers were not replaced by Digital Computers, as the Analog Computers were working fine. I like to think of Analog as Digital with an infinitely fast clock.

OPAMPS IN AUDIO

Opamps are schematically drawn as single Gain Stages with two inputs, and one output. One of their specifications is called Gain Bandwidth Product. It is the Product of the Gain you need, multiplied by the Bandwidth desired. If you want a Gain of 20, with a Bandwidth of 20,000 cycles, you need an Opamp with a Gain Bandwidth Product of 20 x 20,000, or 400,000 Hz (.4MHz). One of the original Integrated Circuit Opamps (1970?), the "741" has a GBP of .5 MHz, and would satisfy our requirements.

SLEW RATE & GAIN BANDWIDTH PRODUCT

Slew Rate has been described as "phase shift," but it is not. It is more like a fireman's problem, "How vertical can I turn the hose, and still reach the floor I'm aiming at?" In electronics it's, "How fast (vertical) can I throw a signal, and how high can I throw it?"

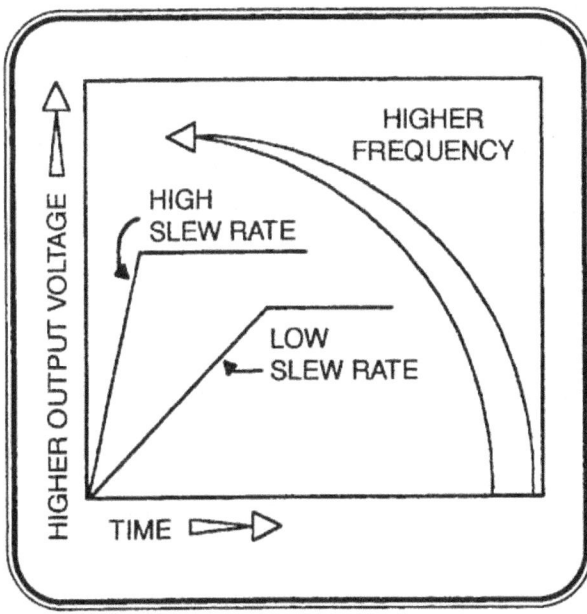

The job of the Opamp, with AC signals, is to throw an Output Voltage to a certain level, at a certain frequency. The higher the Bandwidth, the faster the signal needs to rise. The higher you want to throw it (gain), the faster it has to go to get there in time. Throwing a 5 volt signal, is harder than throwing a 4 volt signal. Throwing a 4 volt signal, 100,000 times a second, is harder that throwing it 10,000 times a second.

At some point, a speed limit is reached. This is the Gain Bandwidth Product, and also the Slew Rate of the device. Lowering the gain, and ultimately, the height of the output signal, allows you to run it faster. Running it faster means you can't get as high, as fast.

"Slew Rate" is how vertical of a line the output can follow. As the output voltage height goes up, or the frequency goes up, the leading edge of that sine wave becomes more vertical. The Slew Rate is defined in Volts per Microsecond. How many volts can the output get to in a microsecond. A Microsecond is .000001 of a second. The Inverse of .000001 is 1,000,000. So, if a device can throw a 10 Volt output signal, 1 million times a second, or 1 Volt, 10 million times a second, it's said to have a Slew Rate of 10V/uS. Due to other factors, the Power Bandwidth of this device is only 200 kHz.

So, Gain Bandwidth Product, and Slew Rate are two ways of saying the same thing. If you exceed the Slew Rate by having too hot, or too high of a frequency input signal, through a stage set for too high of a Gain, you'll develop Transient Intermodulation Distortion, or the dreaded TIM. This is one of the causes of distortion in Solid State devices. It is caused by insufficient Slew Rate. Tubes have less of a problem with Slew rate because they are High Impedance devices. This means they operate on Voltage, and not Current. A Solid State device operates on Current and needs to physically move electrons to operate. Moving electrons takes time, and slows down Transistors.

POWER SUPPLY or OUTPUT CLIPPING

Opamps are usually operated with a Plus and Minus DC Supply of 15 Volts. That means +15 Volts, and -15 Volts. These are also known as the Power Supply Rails. If you drive an Opamp's Output to exceed these voltages, 30 Volts total, you'll get Hard Limiting, Hard Clipping, or Bad Sound. This could be called "Output Clipping."

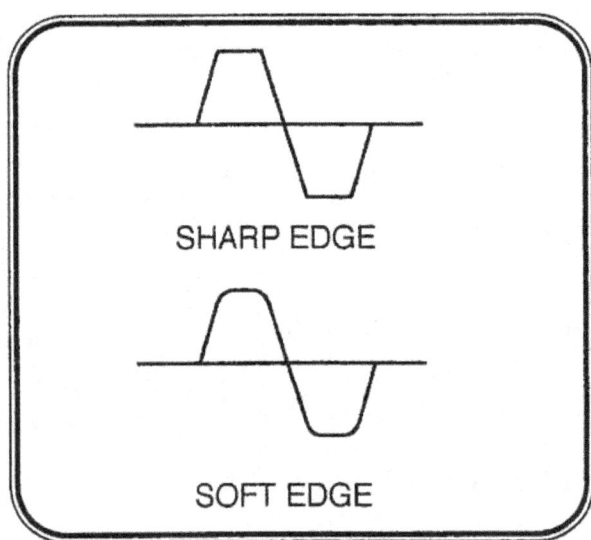

SHARP EDGE

SOFT EDGE

Because of the nature of Hard Clipping, a sharp edge is formed when the Opamp runs out of power supply. The definition of a Square Wave is the sum of all the Odd Harmonics. Odd Harmonics conflict with the fundamental frequency, causing a loss of definition, fullness, and apparent vol-

ume. Power Supply Limiting generates Odd Harmonics, as indicated by the sharp edge of the Wave.

"Soft Edge" distortion contains more Even Order Harmonics, such as the Second Harmonic, that adds beef, and apparent volume to the signal.

So the idea is to Clip the signal, in a way that generates Soft Clipping, and Even Order Harmonics, of unequal Duty Cycle. The best way to do this is with a Four Gain Stage design.

The First Stage powers the "Gain" control, and allows control over the saturation, and the Duty Cycle. It feeds the Second Gain Stage. The Second Stage's output is attenuated by a Resistive Divider, and then goes into the Third Gain Stage. The Third Gain Stage's output is attenuated by a Resistive Divider, and goes to the Fourth Gain Stage.

Notice that the Plate Resistors, Cathode Resistors, Cathode Caps, Coupling Caps, and Divider Resistors, all vary in value. This is to adjust the Gain, and Tone of the distorted signal at each stage, and produce a mix of distortion amounts at different frequencies.

This circuit also has the ability to change the Duty Cycle with variations in how the Gain control is set.

INPUT CLIPPING

You can also clip Opamps by overdriving the inputs, and diodes are provided in most cases to protect the Opamp from this type of abuse, which can destroy it.

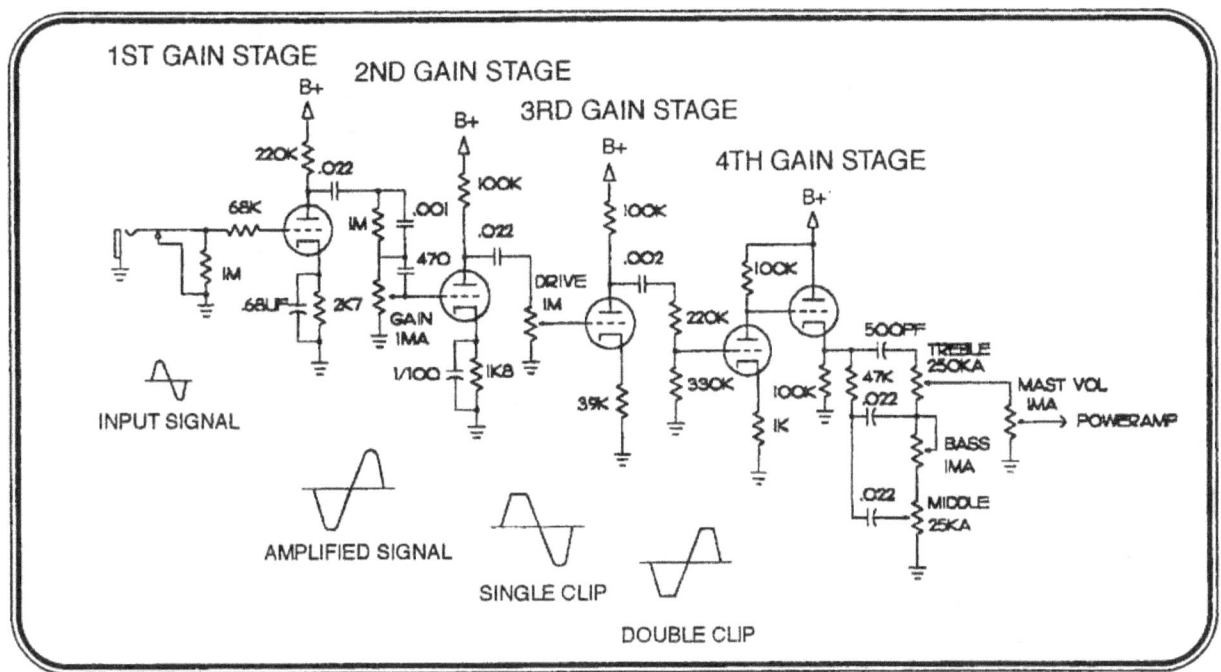

INPUT SIGNAL

AMPLIFIED SIGNAL

SINGLE CLIP

DOUBLE CLIP

Preamp tube distortion is created by over-driving the input of the tube. It can also be called Bias Distortion. Even the hardest driven preamp tube can throw only 70-85 volts of Output. With a 300 Volt Supply, you can see that this is far short of clipping the Supply Rail. But the Grid of a preamp tube can only take the Bias Voltage in before it goes "Positive" relative to the Cathode, and loses linear control of the Plate Current. This is how Preamp Tubes distort.

The reason for multiple Preamp Gain Stages is a tube stage only clips one side of the wave at a time. After the single clipped wave leaves the Plate of the Preamp tube, it is inverted, and the second Gain Stage clips the other side of the wave. Clip one side; invert the signal through a Gain Stage; and clip the other side of the wave. The smoothness comes from a series of stages, each adding a little bit of distortion, and compression. When you do this, the problem is you get that "LA" compressed distortion tone.

If you read that a certain amp has sustain without compression, don't believe it. The reason an amp sustains is because it compresses. You hit the ceiling, and beyond. As the signal drops off, you're still on the ceiling. That's compression, and it doesn't come without excessive gain. Sustain is the effect, as the output remains the same, even as the signal drops off, until you drop off the ceiling. For you guys that are looking for sustain, and then complain about the signal being compressed, that's just the way it is.

An exception to this is "feedback." If you can design the circuit to be locally unstable, you can find a sweet spot where the output couples back to the input, and create sustain through local feedback. This is a difficult balancing act, unless the amp is built on a circuit board, where the parts can be orientated consistently. The Thunderfunk amp uses this little trick, among others.

This multiple clipping process is shown in

105

the drawing on the previous page. The First Gain Stage serves to amplify the input signal, and allows control of the saturation with a Gain Control before hitting the Second Gain Stage. The Second Stage rarely clips, as the output of the First Stage is normally around 4 volts. If the Gain Control is full up, the Second Stage might clip. The Third Gain Stage provides the first solid clipping activity. The Voltage Divider between the Second and Third Gain Stages might be changed to a Control, and could be called "Drive." The Third gain Stage inverts the signal, allowing the Fourth Gain Stage to clip the opposite side of the wave. The Fourth Gain Stage is followed by the Tone Controls, and Tone Recovery amp. At this point there is no need for addition distortion to be added to the signal, and all the gain structure is aimed at providing enough push to the power amp.

I personally believe that the Gain should precede the EQ, although I build both designs. I suspect that a problem is created with the Post-EQ amps in that if you increase the EQ you change the Gain structure, and therefore the Tone. Adjust the Tone to compensate, and you change the Gain structure and Tone again. It seems to be a constant balancing act.

The Trainwreck Express is setup the same way but doesn't have this problem because the Gain Structure is such that the Power Amp distorts before any stage of the Preamp, including the Inverter Driver. The result is pure Power Amp Distortion, closely followed by Preamp Distortion. This adds a smooth flow of distortion turnover and points out the hardest thing to make a distortion amp do; change smoothly from clean to dirty. That edge is where a lot of players like to hang

out. If you can get a touch going, where the dynamics of the players touch changes the texture of the distortion, you'll get everyone excited about your amp. Ken Fischer has done just that with his Trainwrecks. His output transformers have this great mid-range frequency response. Another reason Ken's amps work so well is they are all around 30-watts of power. This allows Trainwrecks to be used for Power Amp Distortion. If I haven't said this before, I'll say it now. If you're not on Ken's waiting list, you should get on it now. It's the only amp I know that ALWAYS sells for more used than new. It's impossible to lose money on the deal, if you ever wanted to part with one.

The Trainwreck is an example of Power Amp distortion. If 20-30 watts of power isn't adequate for your needs, then you'll most likely end up with an amp that can generate Preamp distortion. In most cases, amps which generate Preamp Distortion, can also generate Power Amp Distortion, when the controls are set properly. The advantage of Preamp Distortion is the use of completely saturated tones at low volumes, and the ability to generate clean, dynamic tones at high volumes. A low powered amp can do the first, but is limited in its ability to play loud and clean.

DISTORTION WAVEFORMS

Whether the distortion is generated in the Preamp, or the Power Amp, the desirable characteristics can be seen in the Output Waveform.

When a Sine Wave is clipped, it turns into a Square Wave. A Square Wave contains a High Frequency component, and a Low Frequency component. The vertical sides

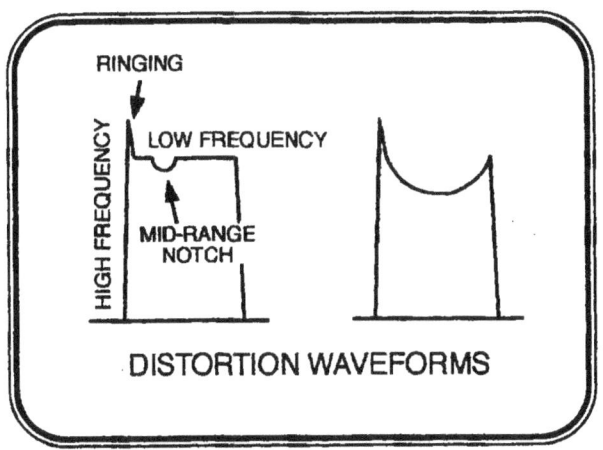

DISTORTION WAVEFORMS

of the Square Wave represent High Frequencies, while the flat top of the wave represents Low Frequencies, or DC for that matter. When you try to run a Square Wave through a circuit, at levels exceeding its linear levels, the Square Wave gets further distorted by the circuits inability to "track" the Waveform.

At the rising front of the Wave, you'll get "overshoot," and "ringing." Across the top of the Wave, you'll get a non-flat response indicating a deficiency at some mid-frequency.

The "ringing" produces an exaggerated High Frequency response that gives the sound a "biting" clarity. Without it, The distortion sounds "dull."

The "dip" in Mid-Frequency response, is what I like to call, "The Marshall Notch." It gives the sound a "hollow howl" that I'm sure you're all familiar with.

Soldano amps have an even more exaggerated "phasey howl" caused by his not turning off his Clean Channel when he turns on his Dirty Channel. The phasey howl is caused by the fact that the two channels are out of phase.

You can generate the howl by turning down your mid-range control, at the expense of losing your sustain. If you're not able to generate a waveform that looks something like those shown, chances are your amp is a dud. Don't be fooled by looking at the output of some distortion foot-pedals, like the Fuzz Face. The Square Wave it generates goes into the amp, where the ring and frequency cancellations occur.

CONCLUSION

Now that we've seen how distortion can be created, let's look at the next stage of an amp, the tone controls.

CHAPTER 15
TONE CONTROLS

Why does a Tweed amp sound like a Tweed amp? Why is it different from a Black face amp? One big reason is the Tone Controls. Equalization is a very important way to modify the sound of an amp. It is far more than just a way to compensate for the tonal differences in bridge and neck position pickups. I'm often amazed by guys asking for the tone on a certain record, never thinking about the racks of Compression, Expansion, Gating, and EQ available in studio recordings. How do you know THAT sound came out of an amp? The Beatles overdrove their TUBE mixing consoles to generate distortion. Now, how do you get your Vox amp to sound like that? EQ can produce a lot of variation in tone. A wimpy amp can come alive with properly applied EQ mods. If you have one of those amps where the Tone Controls don't work, you're stuck with a one trick pony.

ACTIVE OR PASSIVE?

The Tone Controls on your amp are a Primary source of your Tone! Shouldn't it be obvious? There are a few different topologies to accomplish the control of amplifier tone. One factor often discussed is "Active" and "Passive" Tone Controls. These discussions are almost always wrong in their assumptions. In Engineering (and maybe in REAL life as well) there's an expression:

"When you assume something, you make an "ass" of "u" and "me.""

An Active Tone Control, by definition, has Gain. An example of an Active Tone Control is the Presence Control as implemented on most amps. It reduces the amount of High Frequency Negative Feedback, which increases the amount of High Frequency Gain, but also increases the amount of High Frequency Distortion. The Negative Feedback no longer corrects the Amplifier's Distortion at the Higher Frequencies. With all the Hoopla (Loopla?) about disconnecting the Feedback Loop in order to "Improve" the tone, I've never heard the comment, "Of course, your Presence Control won't work anymore."

The Presence Control is inserted in the feedback loop of an amplifier. This allows it to control the amount of Gain the Amplifier has. This is an Active Tone Control. It can boost a Frequency, as well as cut it. It has Gain. It is tuned to reduce the amount of Negative Feedback at High Frequencies. This causes the amp to have more Gain at High Frequencies,

Remember, the definition of Electronics is Gain. Without Gain we'd all be Electrical Engineers, not Electronic Engineers.

The typical Fender, Marshall, or Vox Top Boost Tone Control works by throwing Gain away. Adjusting it to make it easier for a High Frequency to pass through, raises the High Frequencies higher than the other frequencies, making it sound like a High Frequency "boost." It is really a smaller High Frequency "cut." Being

Passive (it has no Gain Stage) it creates a frequency boost by cutting the level of all the frequencies in half, and restoring some of the selected frequencies to create a perception of a "boost." These controls are followed by what is known as the "Tone Recovery Amp." It "recovers" the Gain lost in the Passive Controls that immediately precede it. In a Fender amp it is the Second Gain Stage, attached to the Wiper of the Volume Control.

So, an Active Tone control directly controls the Gain of a circuit allowing it to have both a "boost" or a "cut." A Passive Tone Control can only throw Gain away, but can give the impression of both a boost or a cut.

I often hear, "Oh, your Tone Controls are Active." What is really meant, is they're effective. They have a lot of "travel." They have "action."

TWEED TONE CONTROL

Many Tweed amps use a simple two Caps and a Pot for a single Tone Control.

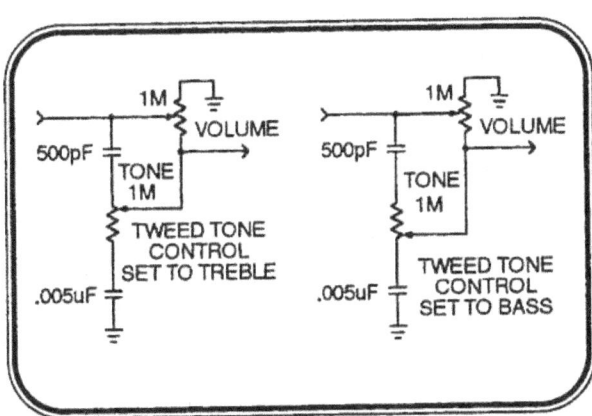

The Tone Control works by allowing High Frequencies to pass through it easier in the high position, and making it harder in

the low position, as well as allowing High Frequencies to bleed off to Ground easier in the low position.

With the Tone Wiper at the top of its rotation, High Frequencies have an easy path through the circuit. You might think of it as an adjustable Bright Switch, putting the 500pF Cap on the Volume Pot at its full up position. At the same time, Low Frequencies are blocked by the 500pF Cap, making its response brighter. Also, the 1 Meg pot is now totally between the Signal and the Cap at the bottom of the Tone Control, putting it electrically far away. In this position, you get a lot of Treble response.

Putting the Tone Control at the bottom of its rotation, now puts a 1 Meg "Resistor" (the Tone Pot) between the 500pF Cap, and the Tone Wiper. Also, in this position the .005 uF Cap is placed from the Wiper to ground. This bottom Cap bleeds off High Frequencies to ground producing a bassier response. This results in a "cut" in High Frequency response. This is about as simple a Tone Control as you can get.

In the higher priced Tweed amps, a separate Treble and Bass control were used. The Treble Control was the same as just described. The Bass Control is a modification of the Treble Control, increasing the Bass response by bleeding off the high frequencies.

TONE CUT

The Tone Cut circuit on Vox amps works by shorting out the High Frequencies as they pass through the Inverter on their way to the Power Tube Grids.

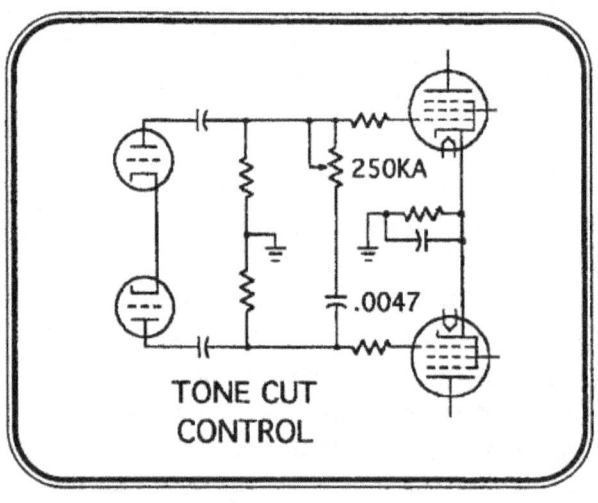

TONE CUT
CONTROL

MODERN FENDER
TONE CONTROLS

This is the most common Tone Control circuit used today. It is found in most Fender, Marshall, and Vox amps. It consists of three Caps, a "slope" resistor, and two or three Pots. The circuit is in balance, and any change to one part of it changes all the other parts. By increasing the "Treble Cap" (the 250pF) you make it easier for high end to get through the circuit (brighter), which also lets more mid-range through. So, have you increased the Treble response, or lowered the frequency range of the Treble Control, or both?

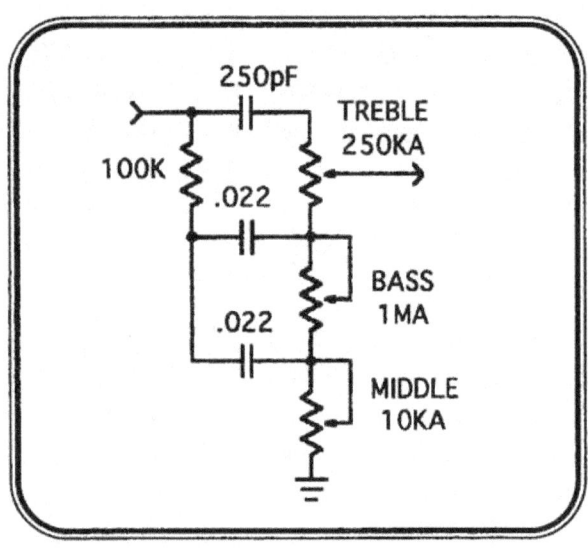

Why does a Tweed amp sound different from a Black face amp? Here's one big reason why. The Black Face, Marshall, and Vox amps have a "modern" Tone Stack which creates a sharp mid-frequency dip centered around 200-400 Hz.

The Tweed amps have a nearly "flat" frequency response when the Tone Control is full up. This gives the Tweeds an accentuated mid-band, which makes the amps sound snappier, and more "Presency." This EQ also contributes to their sustain.

EASY MODS

An easy way to improve the sound of a Fender amp is to replace the .1, and .047 caps found in a Fender with a pair of .022's. The result is a more open, breathy sound. Change the 250pF cap to a Silver Mica for a smoother, less brittle high end. Change the 250pF to a 50pF, and the 250KA Treble Pot to a 1MA for a Vox style Tone Control. The effect is a smoother, slower control of the high end. If you like it, use it.

The 10K Middle Pot shown can be replaced by a 25KA or 50KA Pot for an easy Middle Boost. I use the 50KA Pot since with a 10% Taper, at 1/2 rotation, it has a resistance of 5K. This puts it close to the 6K8 Resistor that Fender uses on non-Middle Control amps, and the 50K value gives you a little more Mid-Boost than the 25K.

There have been other modifications to this circuit by Fender, Marshall, and Vox, especially for Bass amps. I've found that the guitar control works better for Bass than any of the variations.

BRIGHT SWITCH

Did Fender choose a 120pF Bright Switch Cap on the Black and Silver Face amps so there would be no doubt about it working? I can see a kid in a music store saying, "I can't hear a difference with the bright switch," and the sales department saying, "Make it a 120 puffs." I back these down to 100pF, Silver Mica, of course.

BRIGHT BALANCE CAP

You should also change the 10pF Cap across the 3M3 resistor to a Silver Mica. It's used to set the Treble balance of the entire amp, and the ceramic cap has a harsh sound to it. To make the entire amp brighter, increase the size of this Cap.

This is a good place to illustrate the effect of Circuit Impedance on Cap size. Why is this Bright Balance Cap so small (10pF)? It doesn't take much Capacitance to overcome the large 3M3 Resistor. If you made the Resistor smaller, the Cap would have to get larger to keep things in balance.

REVERB MIX

The 3M3 Resistor serves two purposes. It reduces the level of the clean signal before it is re-amplified, and mixed with the Reverb Signal by the Reverb Mix Amplifier. It also provides isolation between the Input to the Reverb Circuit, and its output. So, the Reverb is generated across this Resistor. Any change to this Resistor value will affect the Reverb. I generally call this Resistor, the Reverb Mix Resistor, because the Reverb is Mixed, or Generated across it. More on Reverb later.

TONE CONTROL VALUES

What do the Tone Control values do? What are the differences between driving the Tone Stack with a Low-Impedance Cathode Follower, like Marshall, or driving it with the High-Impedance off the Plate? Can I change the Frequency centers of the controls, and what values do I use to do that? These are all questions I get all the time, and I want to ask you; if I told you a certain value gave you a certain frequency response, what would that mean to you? Do you know what your listening to now? Can you hear the difference between a 250Hz mid-dip and a 300Hz one?

My point is, don't depend on the graphs in this article to tell you what something sounds like. I'm printing a bunch of them to demonstrate what effect the different parts have on these circuits. This is for your education. Learn from the charts, and then go out and listen. You decide what values work for you.

There's an article I want to write, called, "How To Start Your Own Tube Amp Company." One of the pieces of advice I'd give is, "Listen to Everybody, but Don't Listen to Anybody." Take criticism from Everybody, but don't follow what any One person has to say. Take a consensus, and then do what YOU think is right. The Thunderfunk amp uses a 56K Slope Resistor, although I rarely change Fenders to that value. What works with one amp doesn't mean it will work with another. My charts show what effect a certain change has, but does not advise you as to how it will sound with the way you play, or your particular amp.

FENDER CIRCUITS

The first graph compares the difference in response when the "Fender" circuit values of .1/.047 get changed to .022/.022. The larger Fender values move the mid-range dip lower. In doing so, they also decrease the amount of bass response you get, which also means you'll get more high end. I prefer the .022 values.

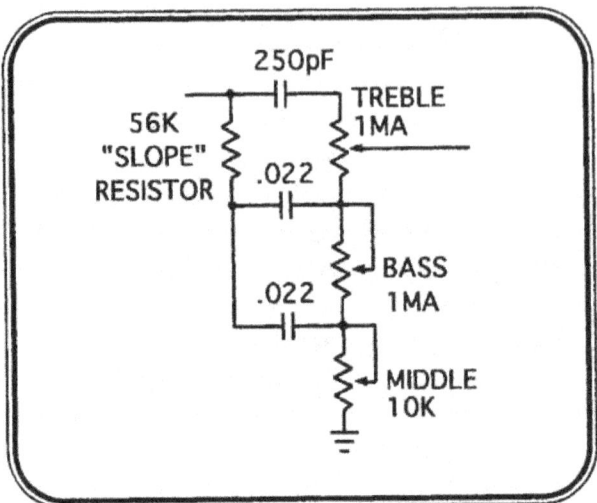

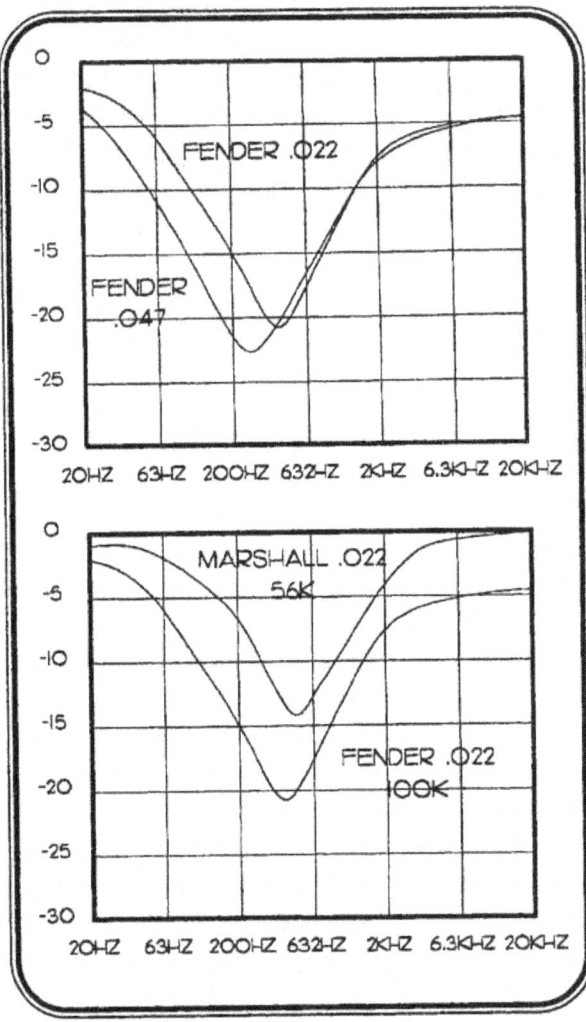

"Slope," or "Divider" Resistor, from 100K with the Fender, to 56K in the Marshall, and the fact that the Marshall supplies the Tone Controls from a Low-Impedance Cathode Follower circuit. The Lower Impedance causes a slight increase in overall response.

VOX CIRCUIT

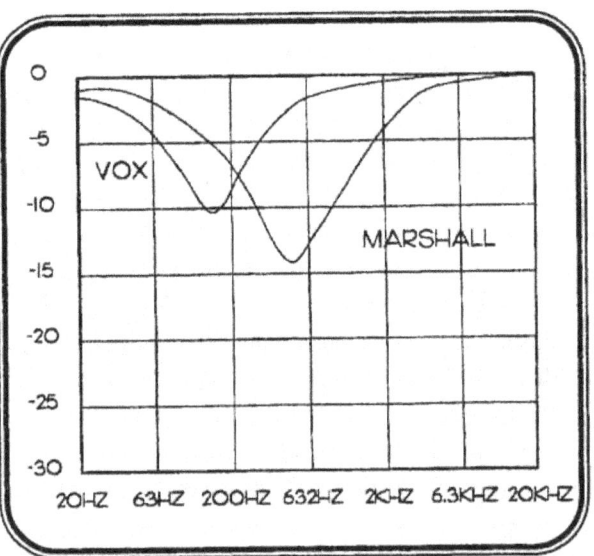

MARSHALL CIRCUIT

The second graph shows the change between a Fender with .022 caps, and a Marshall with .022 caps. The response difference is caused by the change in

Changing the Treble Cap from 250 puffs to 50 puffs, and increasing the Bass Pot to 1 Meg gives the Vox/Marshall comparison. The Vox values are connected to both EL84 tubes in the AC15's and 30's, and EL34's in the AC50's and AC-100's.

It gives a stronger Middle, and Treble response closer to a Tweed style amp.

All Controls are full up, with a fixed 10K Middle Resistor.

SLOPE RESISTOR

The "Slope" Resistor, controls the "Slope" of the Mid-Range dip. It doesn't exactly change the "Q" but rather controls how much the frequencies are divided between the three Capacitors. Think of it as sitting on top of the 10K Middle Resistor. If the Slope Resistor gets smaller, the 10K looks bigger, which is the same as turning up the Middle Control. This causes a reduction in the Middle dip, which "looks" like a change in the "tightness" ("Q," or Quality Factor) of the filter. This lead to the term "slope" resistor.

Radically increasing the Slope Resistor to 470K, strips out much of the low and middle frequencies, leaving the high end accentuated.

DEEP SWITCH

On some Fender Black & Silver Face Bass amps there is a Deep Switch. This switch is "off" in the "on" position, and "on" in the "off" position. Let me explain this. The way the switch works is, it destroys the low end when it is in the "Off" position. When you turn it "On," it is disconnected from the circuit, restoring normal response. Another example of marketing driving an "enhancement." Of course your bass sounds better when that "Deep" switch is "On," because it's no longer destroying the low end! I still always thought the "Normal" channel in these amps sounded better for bass.

CHAPTER 16
REVERB

Reverberation is the repeating of a signal due to the delayed reflections of the signal off of a room's walls. It is often desirable to create a reverberated signal electronically, to add "ambience" to a "dry" signal. It can be used to simulating a "live" (real) room. For example, a radio disc jockey talks close to the microphone to get the strongest possible signal, without background noise bleeding in. This reduces the level of the room's reverberated signal that gets into the microphone, giving his voice a dry, flat, dead sound. In order to restore the "ambience" of a real room, delayed copies of the original (dry) signal are added back in, restoring the illusion of room ambience.

PLATE REVERBS

The highest quality reverb is created by a "plate" reverb system. A large metal plate, as big as 8 feet in length, is driven by a transducer, causing the plate to vibrate. One or more pickups, similar to a phonograph needle, are attached to the plate at locations differing in distance from the source driver. The motion of the plate causes these receivers to vibrate, producing a delayed electrical copy of the input signal. This multitude of delayed copies is then mixed back in to the dry signal, producing a "wet" signal.

ACOUSTIC DELAY REVERBS

A long, usually folded, tunnel is driven by a speaker on one end, with a microphone at the other. The tunnel provides an acoustical delay path for the signal.

SPRING REVERBS

The Hammond Organ company invented the use of springs to create a delay path. Accutronics was acquired by Gibbs Manufacturing, a division of Hammond, devoted to manufacturing spring reverb tanks under the Hammond patent. Today, Accutronics continues to manufacture reverb spring boxes. It is owned by the Morley Company, located in Cary, Illinois, in the far Northwest suburbs of Chicago.

In a spring reverb, a transducer shakes a spring which has a receiving transducer on the other end. When the signal hits the end of the spring, it produces a signal, and travels back up the spring in the opposite direction, and returns again. This produces a variety of delayed signals, increasing the total "diffusion" of the signal.

REVERB IMPEDANCE

The reverb transducers are made from coils of wire. Coils have Inductance. Inductance resists high frequencies. If the driver transducer has a high impedance value, it will roll off the high frequencies. Since the Plate Load of most preamp tubes is 50K-100K ohms, a serious loss of high frequencies would occur. To overcome this, a transformer is used to convert the Plate Load impedance from 100KΩ to 8 Ω, the same as a speaker.

So, a tube reverb system is basically the same as a small speaker amplifier.

FENDER STAND-ALONE REVERB BOXES

The Fender stand-alone reverb boxes used 6K6's (the new ones use the very similar 6V6) through a single ended output transformer to drive the spring box. Basically, a Champ amp.

The signal flow through the Fender box is, Preamp, followed by parallel paths of Reverb and Dry Signal. The Reverb and Dry Signals are "Mixed" by the Mix control. The familiar Reverb control on Fender amps is the same as the Mix control on the stand-alone box.

The Dwell control is a drive level control in the Reverb path. It allows you to adjust how hard you drive the spring box.

The stand-alone boxes also have a Tone control. The only amp that provides a tone control on its reverb system is Thunderfunk. It allows you to set the amount of Treble response in the Reverb signal. A bassy reverb goes into the background, while a bright reverb moves forward in the mix.

SOLID STATE DRIVE

If you try to drive an 8 ohm spring box with a opamp, you'll burn it up. You could use the opamp to drive external power transistors, which could supply the Current necessary to drive an 8 ohm box. Power transistors are expensive. An easier solution is to raise the Impedance of the spring box to a level that the opamp can drive easily. The resulting roll off of

high frequencies can be compensated for by a filter, placed before the reverb driver, to increase the amount of drive at the higher frequencies. If not properly designed, this circuit could have increased distortion at high frequencies, or an excessive fall off of high frequencies.

REVERB BOX TYPES

The Accutronics company uses a coded part number to describe their spring boxes. The first four letters or numbers are as follows:

The first digit identifies the basic type of box.

TYPE 1   A small, 2-spring box.

TYPE 3   Special purpose 4-spring box without shock mounting.

TYPE 4   The old standard Fender 4-spring box.

TYPE 6   A very small, shock mounted, 2-spring box.

TYPE 8   A small, 3-spring box.

TYPE 9   The new standard Fender 6-spring box.

The second and third digits identify the box's Input and Output Impedances. The meaning of these Letters varies with the different family of boxes.

So far, we've covered the first three letters of the Accutronics reverb part number. It gets complicated, so I've reprinted the entire Accutronics Part Number Chart. You can refer to these charts and have an easy "look up" to for replacement parts. The remaining digits are as follows: The fourth digit, "2" denotes the "decay" of the spring box. The number 1 is a short decay for vocals; number 2, is medium for guitar, and number 3 is long, for organs.

The fifth digit denotes the type of connectors. Fender used reverb boxes that had the Input Connectors insulated, and the Output Connectors grounded.

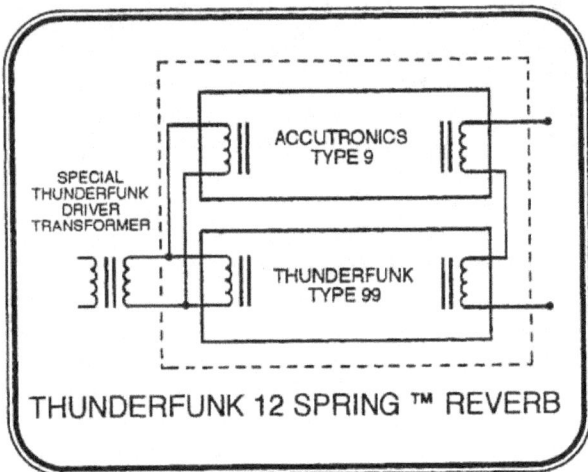

THUNDERFUNK 12 SPRING ™ REVERB

A Thunderfunk 12-Spring Reverb needs an insulated output connector, so the coils from the two boxes can be in series. If they were in parallel, they would phase cancel. I also need a Grounded Input Connector.

The main purpose of grounding one of the connectors is to ground the reverb chassis to the amp chassis. This creates a shield to reduce noise in the reverb system. By the way, I've found reverb systems to be the noisiest part of an amplifier's circuit. It's a real struggle to get it quiet.

The sixth digit tells you whether or not there is a spring lock on the unit. A lock prevents the springs from vibrating when you travel.

The final digit is for how the box is suppose to be mounted. Designers might stick a box in the back of an organ, or across the bottom of an amp. When the reverb spring box is built, the springs are centered in the coils, and that operation depends on which way gravity is going to be deflecting the springs.

REVERB DRIVE AND MIX

In a Fender amp, the signal that gets sent to the Reverb is taken off the Tone Recovery Amp. A 500pF cap connects the tubes output to the Input Grid of the Reverb Driver tube (12AT7).

This signal also goes to a 3M3 resistor with a 10pF across it. The 3M3 resistor attaches to the top of a 220K. It should be obvious that this is a resistive divider. It's the equivalent of a Pot set to a low number. This throws away a lot of the "Dry" (non-reverb) signal, that is recovered in the Mix stage.

The signal that goes through the reverb is recovered by the Reverb Recovery Amp, and its output is mixed in, through a 470K resistor. This blended signal then goes to the Reverb Mix Amp which restores the gain that was thrown away by the 3M3 resistor.

The amount of reverb signal is determined by the Reverb Depth control. It's purpose is to adjust the amount of reverb "mixed" back in to the dry signal. On the stand alone Fender reverb boxes, this control is more properly called "Mix." Its output goes through a 470K resistor, and this is attached to the junction of the 3M3, and 220K resistor. The 3M3 and 470K resistors also serve as isolation resistors, to

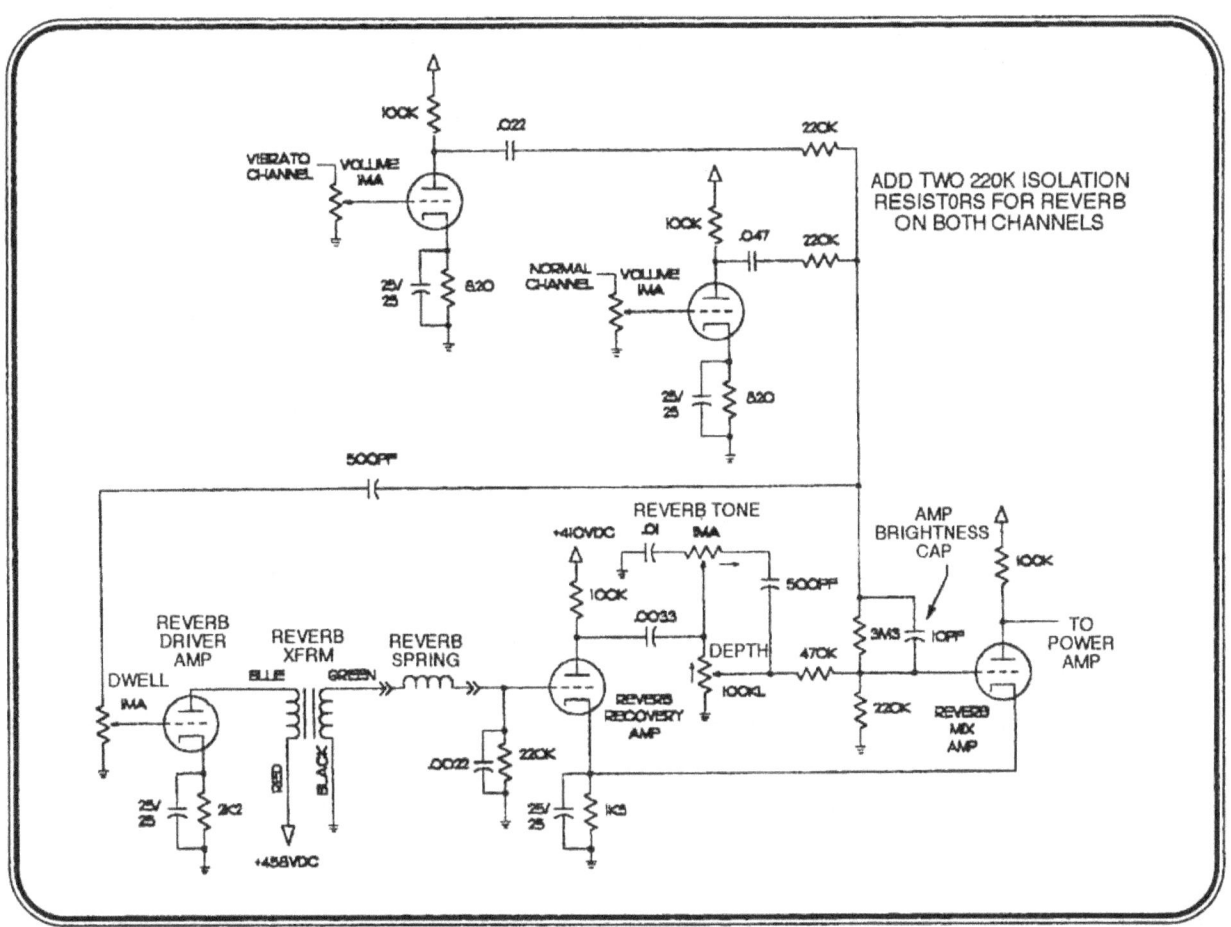

keep the signals from bleeding "backwards" into each other.

There's frequency adjustment going on in the reverb circuit as well. The 10pF cap across the 3M3 serves to increase the brightness of the dry signal, and the overall amp. The 500pF coupling cap to the reverb's input rolls off the low end, to avoid rumble in the reverb spring system. The usual low end roll off point for spring reverbs is 200 Hz. The lows below this point would be blurry and muddy, so they're thrown away. On some amps, there's a small .002 cap across the Reverb Recovery Grid Leak Resistor that rolls off some of the high end off the spring. And finally, there's a .003 Coupling Cap from the to the "Depth" control to further reduce low end coming

through the springs. Since, you've rolled off the low end going into the spring, the low end coming out is most likely noise, like stage rumble.

ADDING A DWELL CONTROL

Changing the Grid Leak resistor on the Reverb Driver tube to a 1 meg pot, gives you a "Dwell" control, like the stand alone boxes. The Dwell control sets the amount of drive sent to the springs, and allows you to get more "Depth" out of the reverb. Basically, the harder you shake the springs, the longer the reverb will last.

So, the Dwell control adds "delay," or more properly "persistence," to the reverb. Since changing the Grid Leak Resistor to a Pot will only allow you to reduce the amount of drive, and not increase it, you might need to add gain to the drive tube to allow more adjustment range. Then you could actually "increase" the amount of Dwell over the standard circuit.

CHANGING REVERB BOXES

I find the old 4-spring boxes to be obnoxiously springy sounding. The 6-spring box is a big improvement. Adding 50% more delay paths has a synergistic effect, and the reverb is much more than 50% improved.

There's some high end equipment that has gone back to the 4-spring box to capture the "vintage" sound. I think this is ridiculous. In the '50's, we wanted our reverbs to sound better. A very simple mod for a Fender amp is to change the Type 4 box to a Type 9. The new part number is 9AB3C1B. Notice that Fender uses the "3" delay, which is the longest. I don't know if this has always been their standard, but I find the long delay to be too much. The short delay is appropriate for vocals, in that it keeps the reverb clearer. I find it too dry though, even for vocals. I use the medium delay box for my designs.

THUINDERFUNK
12 SPRING ™ REVERB

The Thunderfunk 12 Spring Reverb uses a standard Accutronics Type-9 Six Spring reverb box, in combination with a custom made Thunderfunk Type-99 spring box. The Type 99 has different delays designed into it to compliment the Type 9's. Just like going from 4-springs to 6-springs, the new 12-Spring Reverb is more than twice as good as the 6-spring. I'd place it as

95% as good as a plate. As far as I know, Thunderfunk is the only company making a reverb of this quality.

If you look at a spring box, you'll see two springs joined in the middle by a rivet. Notice that the rivet is not in the middle, nor is the rivet in the adjacent set of springs, in the same place. When the spring vibrations reach the mass of the rivet, some of the energy is reflected backwards, causing addition diffusion. The springs in a reverb box all have different lengths, wire gauge, and tension. Accutronics designs the springs to be different in order to increase the amount of diffusion created by the box. My attitude about modding the reverb tanks is, "Accutronics knows more about reverb than I ever will." The same thing with transformers. I use transformer engineers, who are a breed unto themselves. I'll never know as much as these guys do about their specialty. My advice is, don't mess with the insides of the tank.

BROKEN REVERB

While this is a descriptive article about reverb, and not a troubleshooting article, a few simple tips are appropriate here. The tiny wires from the RCA pin phono connectors to the reverb coils, like to break. If your reverb doesn't work, first check the continuity of the braided reverb wires from the amplifier to the reverb pan. If they're fine, then remove the spring box and visually inspect the wires inside the box. If they're broken from the connector - reattach them. If they're broken from the coil - throw it out. If it's a valuable vintage piece, try calling the factory about a rebuild. Most reverb problems are in the wiring, or the tubes.

REVERB RECOVERY AMP

The Reverb Recovery Amp amplifies the low level signal generated by the reverb spring receiving coil. Not much to change here, except to reduce feedback and noise in a reverb system, you might try reducing the gain (slightly) on this stage. The idea is to amplify less signal, and also less noise. Then increase the amount of drive to the spring, which should restore the amount of signal lost by the gain reduction. Keep these changes minor, as the circuit is sensitive to drive and gain levels, and part values are pretty much tweaked already.

REVERB TONE CONTROL

An easy mod at this point is a Tone Control. I designed this circuit by blending a standalone box design, and a Champ's tone control. It's pretty fundamental, but effective. By adding a Dwell control, and this Tone Control to a single knob Fender reverb system, and you'll have all the controls of a standalone box. Thunderfunk amps have the Tone control, but not the Dwell. So, in order of importance, I'd say it's Depth (Mix), Tone, and Dwell.

GAIN CHANGES

Seeing that the 3M3 resistor is a resistive divider in conjunction with the 220K resistor, it's easy to increase the amount of gain through a Fender reverb amp, by making the 3M3 resistor smaller. As I've said before, Fender amps are already tweaked for part values. The way to improve an amp is usually done by changing the circuit, not just values. Of course, the exception is the .022 tone control change, but even that is an older Fender design. My advice - leave the 3M3 alone.

REVERB & VIBRATO ON BOTH CHANNELS

I've seen amps modified to have reverb on both channels, and frankly, they don't work very well. While working on an amp that had this mod, I noticed what the problem was. The channel isolation resistors were not moved to a location before the Reverb. The purpose they serve down stream, right before the Driver/Inverter tube, is to mix the two channels together across these two resistors. If you want to mix the two channels before the reverb, then you'll have to move these resistors as well. These resistors keep the outputs of the two channels from talking into each other's Outputs. The fix became a simple move, or the addition of new 220K Channel Isolation Resistors.

Be aware that combining the channels at this point will also put the Vibrato on both channels. A bummer if you're using the Vibrato on your guitar channel, and singing through the other one. Yeah, you'll have reverb on your voice, but the Vibrato will sound a little strange.

The same warning applies to the amount of gain lost by the Vibrato circuit. This gain loss will now be in both channels.

ACCUTRONIC REVERB PART NUMBER CHART

Type 1 - 2 Springs	A - 8 ohms	A - 500 ohms	1 - Short	A-Input Grounded Output Grounded	1 - No Lock	A - Horizontal Open Side Up
Type 3 - 4 Springs	B - 150 ohms	B-2250 ohms	2 - Medium	B-Input Grounded Output Isulated	2 - Pin Lock	B - Horizontal Open Side Down
Type 4 - 4 Springs	C - 200 ohms	C-10K ohms	3 - Long	C-Input Insulated Output Grounded	3 - Wire Form Lock	C - Vertical Connectors Up
Type 6 - 2 Springs	D - 250 ohms			D-Input Insulated Output Insulated	Options 2 & 3 available for Type 4 only.	D - Vertical Connectors Down
	E - 600 ohms			E - 10" Leads No Outer Channel		E - On End Input Up
	F - 1475 ohms			F - 3" Leads No Outer Channel		F - On End Output Up

Type 8 - 3 Springs	A - 10 ohms	A - 500 ohms	1 - Short	A - Input Grounded Output Grounded	1 - No Lock	A - Horizontal Open Side Up
Type 9 - 6 Springs	B - 190 ohms	B - 2250ohms	2 - Medium	B - Input Grounded Output Insulated	2 - Pin Lock	B - Horizontal Open Side Down
	C - 240 ohms	C - 10K ohms	3 - Long	C - Input Insulated Output Grounded	3 - Wire Form Lock	C - Vertical Connectors Up
	D - 310 ohms			D - Input Insulated Output Insulated		D - Vertical Connectors Down
	E - 600 ohms			E - 10" Leads No Outer Channel		E - On End Input Up
	F - 1475 ohms			F - 3" Leads No Outer Channel		F - On End Output Up

CHAPTER 17
EFFECT LOOPS & BALANCED LINES

When you buy an amp, it might come with reverb and vibrato. Sometimes an amp has compression, or channel switching built in for distortion. With these effects, the designer has taken care of any noise, level or impedance matching problems.

Pedal effects operate at the level of the instrument (Instrument Level). The guitar plugs into the pedal, and the signal coming out is usually no more than three times louder, if at all. Most pedals provide a Level control to match the straight through level, to the level with the effect turned on. When adjusted, the output of the pedal is basically the same as the guitar level going into it. When pedals are strung together, level, impedance matching, and noise all become important factors.

One of the first suggestions for a string of pedals, is to add a "buffer" before the pedals. We've covered buffers before, but here's a review.

BUFFERS

A buffer is an amplifier that has a High Input Impedance, and a Low Output Impedance. The Low Output Impedance can provide a high Current signal to any stage following it. To explain this, let's jump to digital circuits and a problem called Fan Out.

FAN OUT

Digital circuits, or "Gates," are basically switches, just like the ones that turn on your lights. If you have a light switch, and it turns on one light bulb, you're O.K.. If you attach two light bulbs, now the switch has to "carry" twice as much Current to light both bulbs. Continue to add light bulbs, and eventually, you'll have problems. The reason is, every time you add a bulb to the switch, its "Load impedance" is reduced, and the circuit draws more and more Current through the switch and the wire. Eventually, the wire supplying the lights will overheat, and melt. To avoid this, a Fuse is added to the wire. This keeps you from making dumb mistakes, like hooking up too many lights to one circuit. This Fuse is sized to prevent you from starting a fire inside your walls by connecting too many light bulbs.

In digital circuits, designers have the same problem. One digital Gate's Output (the switch), might be connected to multiple Inputs (the light bulbs), all in parallel. Connect more Inputs, and pretty soon, the Gate is driving so many Inputs, and it can't keep up with the Current demand. If the supply of Current fails, the Voltage drops to bring Ohm's Law back into balance. The Current times the Resistance has to equal the Voltage. If the Resistance stays the same, but the Current drops, the Voltage will drop. The measurement of how many Digital Inputs one Digital Output can drive, is called Fan Out.

This ability to "drive" ever lower Impedances is what a Low Output Impedance is all about. It doesn't matter if you connect one light bulb, or more. If the wire is big enough, the Source wire can supply Current without excessive Voltage drop. You can see that if the Voltage drops, that's distortion.

There's also Fan Out in a Graphic EQ. The signal coming in goes to multiple filters, each one tuned to a different frequency, and all connected to the input in parallel. Everytime you attach another frequency filter, the total Input Impedance is being divided down. If you look inside a Graphic EQ you'll see that the input signal first goes through a single amplifier, called the "Buffer." The Buffer isolates the device supplying the signal from all the parallel filters inside. It is designed to provide a High Input Impedance to the Source, and a Low Output Impedance to the filters.

BRIDGING

Voltmeters have High Input Impedances of 1M or more. This allows the meter to "Bridge" signals. In other words, you can attach a Voltmeter to a Voltage without changing the Voltage being measured. If the meter had a Low Input Impedance, it might "pull down" the Voltage being measured, giving a false reading. Add enough Meters in parallel, and even their High Input Impedance gets divided down to the point where the meters will affect the Voltage being measured. This changing of the signal being measured is called distortion.

SOURCE IMPEDANCE

Output Impedance is also known as Source Impedance. Everytime you connect an Output to an Input, you have an Impedance match. It might be good, or bad. When dealing with power, the

Maximum Power Transfer Theory states that the maximum transfer of power between an Input and an Output occurs when the Output Impedance matches the Input Impedance.

This is seen when attaching speakers to an amplifier. When the Output Impedance of a Transformer is 8 Ohms, the amplifier will generate its maximum Output Power when an 8 Ohm Speaker is attached. If you attach a 16 Ohm Speaker, the Transformer won't generate as high of an Output Voltage. Attach a 4 Ohm Speaker, and the Transformer won't generate as high of an Output Voltage, AND the Output Waveform will be distorted.

This distortion is one of the reasons you're looking for an Effects Loop with a Low Output Impedance. To understand how this works, consider the following.

LOAD RESISTANCE

Remember our light switch? Everytime you connect a light bulb to it, the light bulbs filament is in parallel with all the other filaments attached to the same line. Let's say the light bulb has a filament resistance of 100 ohms. Add another bulb, and the Total Resistance equals $1/R1 + 1/R2 = 1/RT$ ohms, or $1/100 + 1/100 = 1/.02 = 50$ ohms. If you take 120 Volts divided by 100 Ohms you get 1.2 Amps. Add the second bulb, and you get 120 Volts divided by 50 Ohms equals 2.4 amps. Add enough bulbs, and you'll draw so much Current that the Voltage fails.

What happens is you have a few Ohms of Resistance in the wire, and that Resistance builds up heat as voltage is dropped across it. From previous articles, you'll remember the Voltage Drop Formula, I * R = E. The more Current you pull through the wire, the greater the Voltage Drop is across it. Wattage is Volts multiplied by Amps. More Voltage dropped across the wire, multiplied by more Amps being drawn through the wire, equals more Wattage being dissipated by the wire. Pull enough Current, and you'll generate enough heat to melt the insulation off the wire. Pull more Current, and the wire itself will melt. This brings us to PA systems.

PA SYSTEMS

Ever wonder why American speakers are 8 Ohms, while English speakers and PA systems are 16 Ohms? PA systems have long runs of cable to the speaker cabinets. Lets say there's 1 Ohm of Resistance in each wire to the cabinet. Two wires, a Positive and a Negative, equals 2 Ohms of wire going to each cabinet. If the speaker is 4 Ohms, the Load on the Amplifier is 4 Ohms plus 2 Ohms, for a total of 6 Ohms. One third of the Amplifier's Power is now being dissipated in the 2 Ohms of wire going to the speaker cabinet.

$$2\,\Omega\,/\,(4\,\Omega + 2\,\Omega) = 2/6 = 1/3$$

There's two ways to handle this. Make the wire bigger, reducing its Resistance, or increase the Impedance of the speaker cabinet, reducing the amount of Current going through the wire, and therefore the Voltage Drop across it.

America is the land of the V8 engine. England is much more sensible when it comes to natural resources. A Marshall amp, with a 16 Ohm cabinet, would have only 1/9 of its power dissipated in the 2 Ohm wire.

$$2\,\Omega\,/\,(16\,\Omega + 2\,\Omega) = 2/18 = 1/9$$

It would also generate a higher SPL (Sound Pressure Level) with the same size amplifier, having less power wasted in the wire.

In America, the answer would be to use a larger gauge wire that might have only 1/2 Ohm of resistance per wire. A few problems with this solution are the wire uses more copper, is heavier, takes more storage space in the truck, and costs more to buy and ship. Using a larger Impedance speaker cabinet saves copper, money, weight, and over the life of the equipment, shipping expenses.

Now, this doesn't really matter in a Fender Twin, when the speaker wire is about one foot long, but Marshalls were designed to be more like a PA system, for guitar. You should now understand that power can be lost in "Source Impedance."

SOURCE IMPEDANCE
LOAD IMPEDANCE

The terms Source Impedance and Output Impedance are the same thing. Load Impedance and Input Impedance are also the same. The Output Device needs to know what the Input Impedance of the following stage is because it's "loaded" by it. Therefore, the Input Impedance of the following device is the Load Impedance of the stage "sourcing" it.

EFFECT LOOP IMPEDANCE

The idea in an Effects Loop is to have a Low Output Impedance, and a High Input Impedance on the Return. The Low Output Impedance does two things. It avoids distortion when the Effects Send is heavily loaded by a device whose Input Impedance is too low. It also allows long cable runs with less noise and high frequency loss. The High Input Impedance allows the Effects Return to bridge the signal coming back from the outboard effect.

NOISE

I don't know if this is technically correct, but I've always thought of the noise problem as one of stray electrons. A High Impedance circuit means that there's a high resistance to the flow of Current. This allows a Voltage to be built up, but that Voltage is carried by a small number of Electrons. A few stray Electrons add a high percentage of noise to the signal.

A Low Impedance circuit is based on Current, and allows Electrons to flow. A few stray Electrons have little effect on the vast flow of Electrons in a Low Impedance circuit.

This is an easy way to visualize how noise affects signals of different Impedances.

CABLE LENGTH & HIGH FREQUENCY LOSS

The cables that connect your effects to your Effects Loop have high frequency loss if they're too long. Twenty feet is the usual maximum length for a High Impedance line. The reason is the cable has Capacitance between its center conductor, and its shield wrapping. This Capacitance Loads the High Output Impedance and causes high frequency loss. This is another reason for a Low Output Impedance.

LOAD DISTORTION

Everything we've covered so far, will help you understand Load Distortion.

If the Output Impedance of the Effects Loop is 300 Ohms, and the Input Impedance of the Effect is 3K, everything is fine. You're transferring a Signal Voltage, and the 3K Input Impedance "bridges" the 300 Ohms of Output Impedance, meaning it takes in a signal voltage without affecting its waveform. This is a mismatch according to the Maximum Power Transfer Theory, but we're transferring a Signal Voltage, not Power. Remember, Power is based on Current. Power amps deliver Current. Preamps deliver Signal Voltages.

If your Output Impedance is 50K, and your Input Impedance is 10K, you're in trouble. This is the reason. When there's no signal, everything is fine. As the signal starts to go up, more and more Current is drawn by the 10K Load (the Effects In). This Current being demanded by the Input, is being pulled through the 50K Source (the Effects Loop Out). This causes an excessive Voltage Drop to develop across the Source, Current Limiting the Signal Voltage. This Current Limiting means that the Effects Loop can't supply enough Current to generate the proper Signal Voltage across the Effects Input. This results in a compressed, if not seriously distorted signal.

Think of the Source Impedance as the wire Resistance in our light bulb example. As you add more and more light bulbs, more Voltage is "lost" across the wire (the Source), and eventually, the Voltage at the end of the wire will be less than the Voltage at the beginning of the wire. This is known as a "Brown Out." If the customers demand more power than the wire can deliver, the result is the Voltage drops to a level that balances the available Current to the Resistance of the Load. Ohms law stays in balance.

I'll leave you with one more example of Impedance Matching, Power Transfer, Bridging, Voltage Drop, and Load Impedance.

I was testing batteries in an environmental furnace at Honeywell. The idea was, we were going to see how long the batteries would backup the alarm systems we were building, when the battery was under load, and in the heat of a fire. The batteries were placed in the furnace, and their temperature was raised. Two large wires were attached to the batteries and ran outside the furnace to large Load Resistors. These were sized to duplicate the alarm system's Current draw. Then two small wires, called "Sense Leads" were attached to the batteries and run out of the furnace to Volt Meters. These were used to monitor the batteries' performance over time and at temperature in the furnace. The question is, "Why run Sense Leads? Why not just use the two large wires, and measure the Voltage at the Resistors?

Here's the answer. The Voltage at the Source (the batteries) is higher than the Voltage at the Load Resistors. The Current that flows through the Load

Resistors also flows through the Battery Wires. The Wires have a certain Resistance to them. This Resistance, multiplied by the Current flow through the Wires, creates a Voltage Drop.

Remember our formula, IR=E (Amps multiplied by Ohms equals Voltage, also known as Ohm's Law). The Sense Wires are much smaller, and therefore have a higher Resistance to them, BUT, the only Current flowing through them is to the Meter, which has a High "Bridging" Input Impedance (1M or more). These Sense Wires allow the Meter to measure the Voltage at the Batteries inside the furnace.

MAKING EFFECTS WORK

So, the idea is to have the effect accurately measure the Signal Voltage at the Effects Output. This allows the effect to get its best "copy" of the Signal.

This is accomplished by designing the effect to have a High Input Impedance, just like our volt meter does. Then the Effect can "read" the Signal without "Loading" it down, causing it to distort from lack of Drive Level.

The other way to solve the problem is to have a Low Output Impedance. This type of an Effect Loop can Drive a High or Low Input Impedance device. It can also Drive a long cable length.

CABLE LENGTH

A guitar player I know was using 2-100 watt tube amps, in a band with a Steel player using a 1,000 solid state amp. He said he couldn't keep up with the volume. Turns out he had a 25 foot guitar cord going to his effects pedals, and another 25 foot guitar cord going to his amps. His problem wasn't amplifier watts. He couldn't keep up because he was losing so much signal in the cords.

The solution was to put a Line Driver at the guitar. This is also called a "Buffer." A Buffer presents a High Input Impedance to the Guitar, and a Low Output Impedance to the Cable, and subsequently the Amplifier's Input. It's not for Voltage increase, since Buffers are normally Unity Gain (1 Volt in, 1 Volt Out). It's for Impedance Transformation. It converts the guitars High Output Impedance to a Low Output Impedance. Impedance Transformation is also the domain of Transformers (Transformationators?).

TRANSFORMERS

An example of a Low Output Impedance Effects Port is the Reverb Send. The Reverb Transformer allows the connection of an 8 Ohm Reverb Box to the High Impedance Plate of a 12AT7. You could plug Effects into the Reverb Loop if you match levels correctly. I've never done

this, or worked it out, so be creative yourself.

HIGH IMPEDANCE EFFECTS LOOP

The simplest way to put in an Effects Loop is to take the Effects Send signal from the Inverter Input Cap, and Return the Loop back to the Inverter. This doesn't work well when driving Low Input Impedance Effects. One of the problems is that 3K3 is considered High Impedance in Transistor circuits, while in tubes it's 50K. If your Effect has an Input Impedance of at least 50K (it will say in your owner's manual), you should be fine. Remember to keep your cable lengths as short as possible, and in no case be over 20 feet.

Coming back from the effect works O.K. because the Inverter Input Impedance is usually 330K or more. This is easily driven by almost any Solid State device. The only other problem is to match Levels. You can have too much, or too little signal going out to the effect, or coming back from the effect. Many effects have adjust-

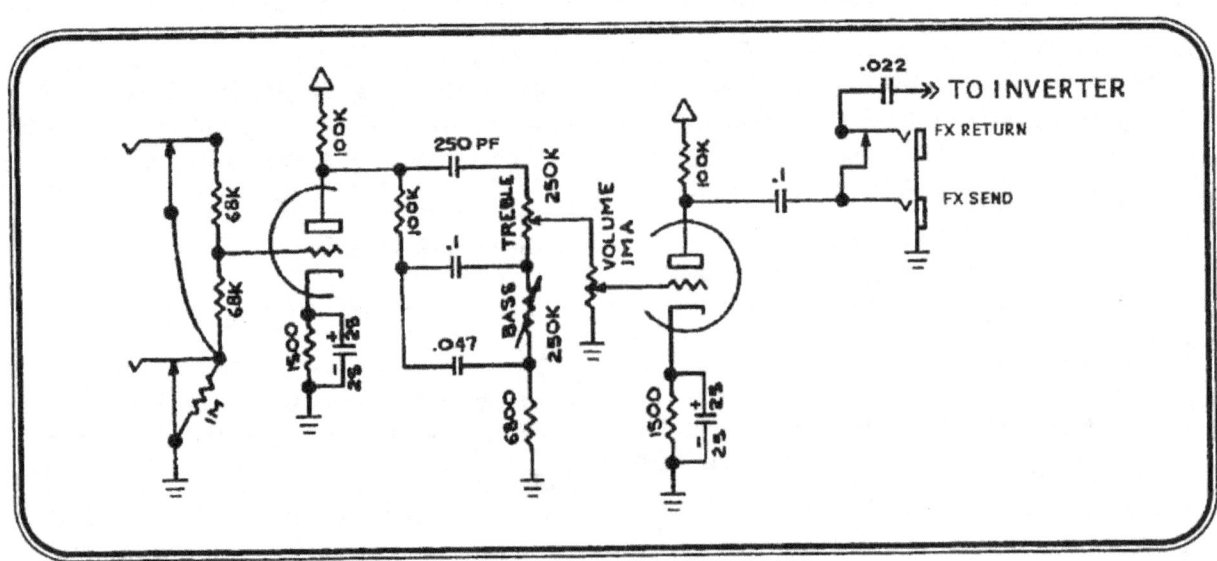

ments on them for Send and Receive Levels. If they don't, you'll have to rig something yourself.

ADJUSTABLE LEVELS

Adding a Pot on the Output and Input, can be used to adjust the Effects Loop Gain. This might work fine, but if the Drive Level is reduced to bring it down to Instrument Level (-16dB), and the Effect returns the same level, there won't be enough Gain to Drive the Power Amp's Inverter. The solution for this is to provide another Gain Stage that can be used to boost the Level of the Effects Loop Return. One Triode stage will do this nicely. If you add a single 12AX7 to do this, it contains two Triodes. What can you do with the other Triode?

CATHODE FOLLOWERS

All the tube stages we've talked about so far, have been Plate Loaded, or Grounded Cathode. The small Resistor from the Cathode to Ground is very small (typically 1K5), and is there for Biasing reasons. These stages will still work if the Cathode is actually connected directly to Ground. Common Cathode stages have a Plate Resistor; the Signal is taken off the Plate;

they have a High Output Impedance; and they Invert the Signal.

If you can ground the Cathode, why can't you ground the Plate? Well, you can. Instead of attaching it to ground, however, it's attached directly to the Power Supply.

While we're on it, there's also a Grounded Grid configuration. It's used in the Long Tailed Inverter, common in most guitar tube amps. We'll cover this in the chapter about Phase Splitters and Inverters.

In the Grounded Plate configuration, there's no Plate Resistor, but instead, there's a large, usually 100K, Cathode Resistor. The Signal is taken off the Cathode, at the top of the Cathode Resistor. This circuit has a Low Output Impedance, and does not invert the Signal. An Output Coupling Cap is needed, since a part of the large DC Voltage that normally appears on the Plate (and still does), also appears on the Cathode.

LOW OUTPUT IMPEDANCE EFFECT LOOP

Combining all the advantages of Level control, Low Output Impedance, and High Input Impedance, we have the Effect Loop.

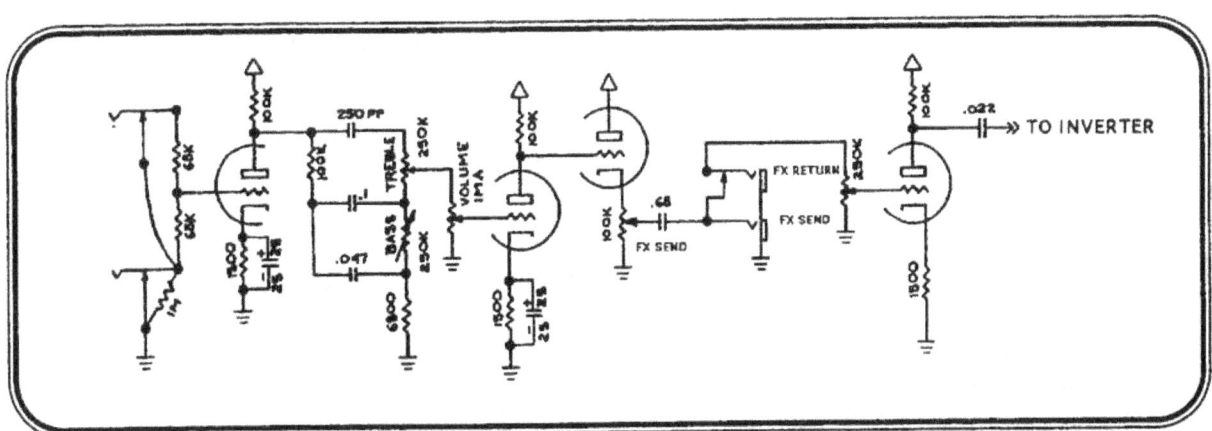

An extra Triode is added to the Reverb Mix Stage, converting it into a Cathode Follower. A second Triode is added to provide a gain stage on the Effects Loop Return. With the addition of Input and Output Controls, a complete Low Output, High Input Impedance, with level setting ability Effects Loop is created.

GAIN SETTING

The loops I use have a Double Pole, Triple Throw Switch to allow setting the Loop to +4dB, -8dB, or -16dB. This allows the Loop to operate with line Level studio Effects (+4dB), as well as foot pedals (-16dB). I don't allow separate adjustment because the user has a perfect opportunity to mess up the gain structure. When you switch the Effects Level, there's no change in amp volume. If the loop is driven harder (+4dB), the Effect Return Gain is reduced to compensate for the expected hotter Return Signal.

The three gain levels are calibrated to produce their stated Levels right at amp clipping. Of course, the Level won't be +4dB if you've got the amp volume set very low.

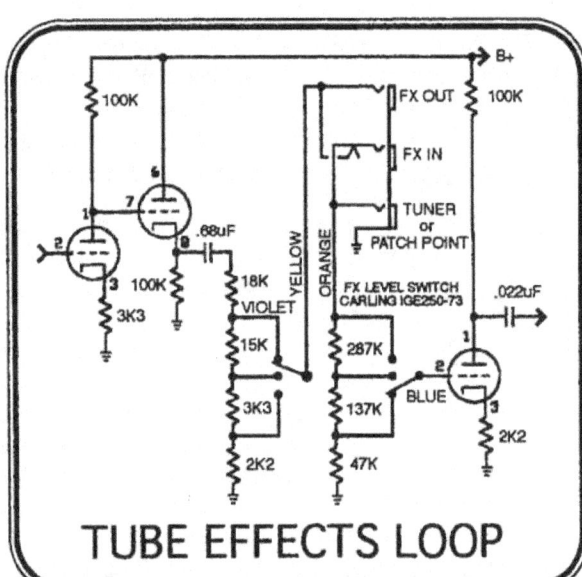

TUBE EFFECTS LOOP

A cheaper way to do this is to use two Pots.

BALANCED AND UNBALANCED LINES

Whenever a signal is sent over a connecting cable, a decision is made about what type of connecting system will be employed. Guitar cables are called Unbalanced lines, and use what's known as a 1/4" Phone Plug, not to be confused with the "RCA Phono" Plug, like the kind used on the Fender reverb system. . Microphone cables, the ones with the three conductor "XLR" type connectors, are called Balanced lines. What's the difference, and why?

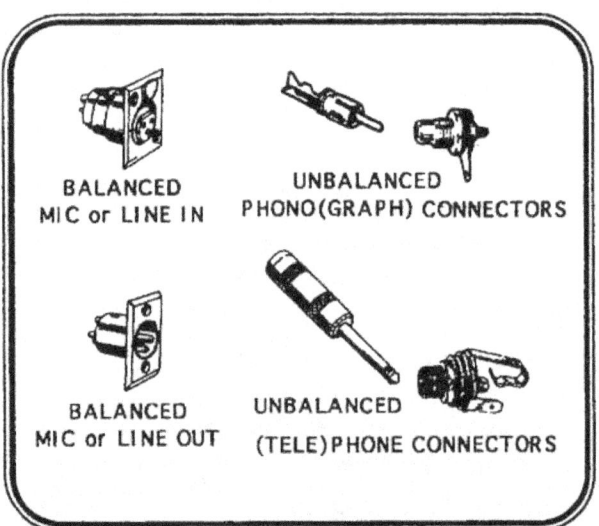

UNBALANCED CONDUCTORS

Electrical signals are transferred from one location to another means of a closed loop. This loop is always two conductors. This allows that for all the energy being sent out, an equal amount of energy is to be returned, with the reverse polarity. Very much like a "push-pull" system.

For every electron being sent out into the line, another electron has to be returned through the "return" line. We're talking about two wires here. One to send the signal out, and another to return the signal. All audio transmission lines work on the same principle. When the Army wants to run a field telephone, they run only one wire. The ground itself is used as a return path. This saves half as much wire as running a two wire system, an important factor during a war. The problem is, this single conductor pretty much acts as an antenna, and picks up all the random electrical fields that are in the air around it, and mixes them back into the original signal. Since one side of this type of transmission line is attached to, and "fixed" at ground potential, and only one wire is actually "hot," we call this type of line "Unbalanced." Unbalanced, that is, with respect to the external "noise" fields. In other words, NOT humbucking.

One way to protect Unbalanced lines from picking up external interferences, is to isolate it from the external fields. This can be done by covering the center conductor with a flexible metal "shield," surrounding it. This shield is normally grounded at only one end, or it might be connected at both ends, or at many points in between. If this shield were to approach 100% effectiveness, it would have to be made of high-quality magnetic shielding material, with an ability to block all electrostatic and magnetic fields as well. Shields made out of aluminum, copper, or other non-magnetic (non ferrous) materials stop almost all electrical interferences caused by electro-static and radio-frequency fields but don't protect the wires from picking up low-frequency magnetic fields from AC transformers, power lines, motors, solenoids, and other sources.

But, with the addition of the Shield conductor, we're back to two wires. The line is still Unbalanced, because the Shield wire is attached to Ground. The Shield, therefore, serves two purposes. First, it covers the signal wire with a conductor at Ground potential, to "drain" off noise fields to Ground, before they can reach the "inner" conductor. Second, it acts as the second conductor for the signal's "Ground Return." The Voltage on the Ground Return wire, in an "Unbalanced" case like this, is 0 Volts, or "Ground."

Since there are no practical shields that can protect an Unbalanced transmission line completely for all types of interference, and shielding a signal cable over a long distance would be very expensive, a lower cost noise reduction system was needed. The phone company's answer to this was the twisted pair system.

TWISTED PAIRS

Basically, two wires are used to carry a signal, and they are twisted together. The theory being that any noise that enters one wire, would also enter the other wire, at approximately the same strength, angle, phase, etc. Remember, the definition of a signal circuit is a closed loop containing a signal wire, and a return circuit. If the signal wire is connected to one of the twisted pair, and the wire of the pair is used as the return circuit, you've got your closed loop. The difference is that the return wire is not connected to Ground.

This type of a "Twisted Pair" circuit is used by the phone company to run long distances, unshielded. It's the same theory as a Humbucking Pickup. In phase noise enters both wires. The signal is applied to both wires, but out of phase (push-pull). The result is, the noise gets phase canceled, while the out of phase signal passes.

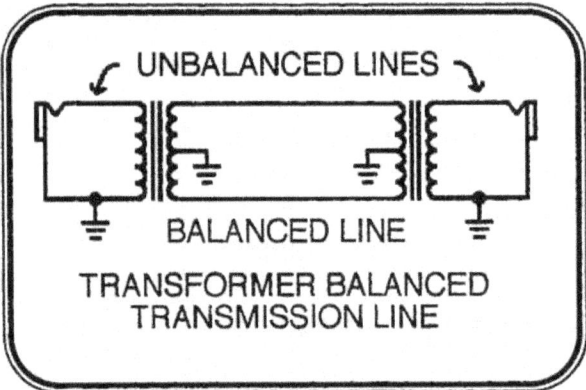

UNBALANCED LINES

BALANCED LINE

TRANSFORMER BALANCED TRANSMISSION LINE

BALANCED LINES

In order to create a balanced line, transformers are used at each end of the line. The Center Tap of at least one of the two transformers is normally connected to Ground. The idea is to keep both sides of the balanced line at the same Voltage relative to Ground. Hopefully, the external fields affecting these lines will affect both sides of the line equally. Although both sides of the line may be saturated with noise from external fields, noise signals arriving at the terminating transformers get canceled out.

How effective the balanced line is depends on two factors. First, the quality of the transformers and the symmetry of its windings. And second, on the symmetry of the interference being induced into the line. The wires are twisted in an attempt to enhance the symmetry of the induced noise.

Balanced lines are cheaper to run for long distances, because they don't require

shielding, are less sensitive to external interference, and their interwire capacitances are lower. They are more expensive because they need transformers, which cost can $30-$75, and up for anything decent, and the transformers themselves can introduce noise, distortion, level, phase shift, and frequency response problems.

The transformers have the advantage of preventing two pieces of equipment from developing Ground Loops between them. In the cases where several pieces of equipment with unknown ground potentials and different power supplies, are being connected, balanced lines are mandatory. This is why you'll find Balanced Lines standard in studio patch bays. You never know what type of equipment you'll need to patch in, and the only sure way to prevent ground loops, noises, melted patch bays, amplifiers, and power supplies is to use balanced lines with isolating transformers.

The transformers also provide an easy way to reverse signal phase. Simply reverse the two signal wires. With an Unbalanced circuit, a transformer or phase inverting amplifier has to be added to accomplish the same thing.

UNBALANCED LINES

Unbalanced lines are more economical to use over short distances, but the cabling cost is expensive for long runs. They require more care in connecting equipment in order to prevent ground loops, are more susceptible to external noise interference, and have high interwire capacitance, which kills the high end response over a cable run of more than a few feet.

WHICH ONE TO USE?

How do you decide which type of transmission line to use? If your Effects are located within ten feet of cable length from your Effects Loop out, then Unbalanced cabling is all right. Remember, to watch for Ground Loops. You might want to isolate you equipment from the rack rails to prevent Chassis Ground connections being made at this point.

If you locate your effects more than 20 feet from your amp, you'll definitely need to run Balanced Lines. This is easily accomplished if you have Balanced connectors in your amp and you effects. Chances are you don't. You can convert Unbalanced lines to Balanced lines by inserting microphone transformers into

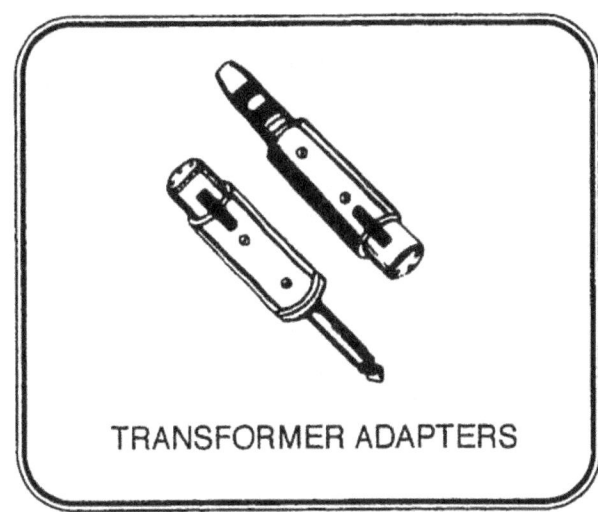

TRANSFORMER ADAPTERS

the line at both ends. These are available from Shure, as well as other manufacturers. You'll be adding the transformers to the system that most companies leave out due to the high expense.

CHAPTER 18
PHASE INVERTERS

When discussing tube amp circuits, the single most asked question I get is, "How does the Phase Inverter work?" The explaination of how they work starts out simple, and then gets kinda complicated. We'll approach it with a lot of common sense.

There are three inverter circuits you'll find used in guitar amps. The Paraphase Inverter was the first used. This evolved into the "Cathodyne," or maybe more obviously, the "Split Load Inverter." The last one is called the "Long-Tailed Pair," "Schmitt Inverter," or most accurately, the "Cathode-Coupled Inverter."

WHY DO I NEED A
PHASE INVERTER?

The first amps were single ended affairs, running Class-A Bias, and having no need for a Phase Inverter. A signal came in to a Class-A preamp circuit, and was passed on to a Class-A power tube. The signal went from "Ground" up. It never went below Ground.

To get more power, we could use two tubes. Let's put them side by side, and double our power. Boy, is it getting hot in here. Those Class-A amps really run hot, don't they. Class-A means that even when there's no signal, the tube is still biased at 100% of full power. It will go up to 200%, and down to 0%, and the average power remains — 100%.

Class-B is nice, because the tubes are biased to be off when there's no signal. Getting much cooler now. But, if the tubes are biased Class-B, you'll need one of them to go from 0% to 100%, and the other one to go from 0% to -100% — a negative number. That's what a Phase Inverter does. It flips the signal upside down to provide the negative "Pull" signal for our new "Push-Pull" circuit, using two tubes now running the much cooler, and more efficient, Class-B.

Maybe you see the problem. In order to go from 0% to -100%, we'll need a Negative high voltage power supply for the bottom tube. No problem, we'll fool the circuit by "floating" it. The primary of the output transformer is not reference to ground. It will allow the power tubes to gyrate around the DC voltage. Did you notice that in opamps, plus and minus supplies are standard? This allows them to output positive or negative voltages.

Since the power tube are "gyrating" around each other, while the top tube is going up, the bottom tube can go down. They "Push" and "Pull" at the same time, in oppostion. This might explain to you how a Vox AC-30 can be Class-A, AND Push-Pull at the same time. The tubes are in opposition, all the time. To what class they're biased is really of no concern to our circuit.

Now, before we go on, let's restate the fact that these Push-Pull amps are in fact biased to Class-AB. This means, that for small signals, both tubes are pushing or pulling at the same time. Big signals cause one of the tubes at a time, to be driven into cut-off.   So while the top tube is cutoff, the bottom tube continues on, and vice versa. Now the circuit is overlapping into Class-B.

Besides an increase in efficiency, the nature of Push-Pull is to cancel even order harmonics in the output stage by the symetrical nature of the circuit.

TRANSFORMER COUPLED INVERTERS

Now all we need is a way to generate an opposite signal. The first way was to use an Interstage Transformer. This is a signal transfomer that goes "between stages," hence its name. To understand how it works, look at the drawing below.

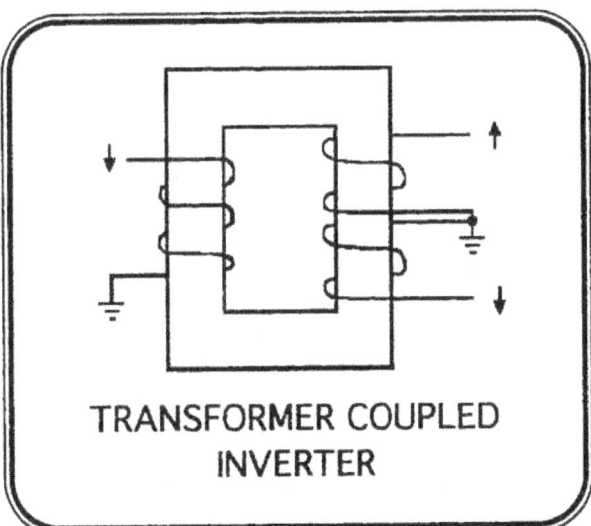

TRANSFORMER COUPLED INVERTER

When a signal travels through the primary, it induces by magnetic coupling, a signal in the secondary. By grounding the bottom of the top secondary, we get a certain phase output that is indicated by the arrows. If we grounded the bottom of the bottom secondary, we'd get the same phase output as the top, but we're ground-

ing the TOP of the bottom secondary. This results in an output that's 180 degrees out of phase with the top signal. Now, tie the two center windings together and attach to Ground, and we have a positive and negative "Push-Pull" signal referenced to ground. If this looks familiar, look at the section on Balanced and Unbalanced Lines.

Transformers are expensive, so engineers designed circuits to accomplish the same thing, electronically.

THE PARAPHASE INVERTER

The first inverter circuit used by Fender in the late '40's and early '50's, was the Paraphase Inverter.

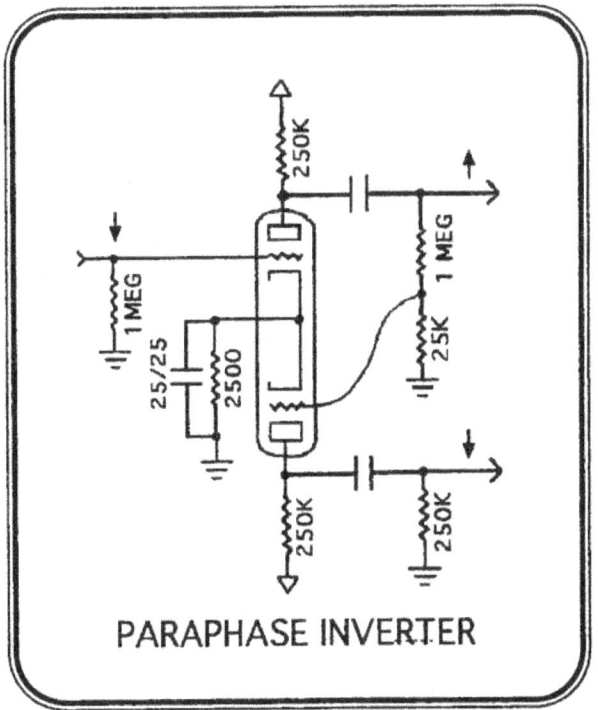

PARAPHASE INVERTER

The input signal is first amplified and inverted by the top inverter tube. You'll remember that when a signal goes into a

Grid, and out of a Plate, its phase is reversed. Some of the top tube's signal is taken off of its output Plate, and returned to the Grid of the bottom inverter tube. Having gone into the Grid of a tube, and coming out of its Plate, its phase is reversed, again. Now we have two signals, each opposing each other.

So, the signal at the top power tube is the inversion of the input signal, and the signal at the bottom tube, having been inverted twice, is back in phase.

In order to balance the signal to the output tubes, the voltage divider ratio of the signal feed resistors should equal the gain of the top inverter tube. For the "5C3" Deluxe, this ratio is 1M/25K = 40. The Amplification of a 6SC7 is 70, making the voltage gain through it an estimated 28 to 42. So, the Fender values may not be perfect. 35K appears to work better. You might want to add a pot and do some balance tests with your own tubes, since they may not have been in balance when the 6SC7 was new, and it certainly has changed with age. Replace the 1M and 25K resistors with a 1M pot, and adjust it for balance or tone. Measure the value of the "lower" side. Replace the 1M resistor, and use the new value for the 25K.

Whether you want the inverter to balance, or not, is something I discuss in the Chapter 12 section on "Tone."

Other problems is the signal to the lower half has been delayed (phase shifted), and its frequency response further distorted by passing through two tubes instead of one. The next improvement was the Self-Balancing Paraphase.

THE SELF-BALANCING PARAPHASE INVERTER

To better balance the inverter, Fender switched to a "Self-Balancing Paraphase Inverter" for the 1954 "5D3" Deluxe.

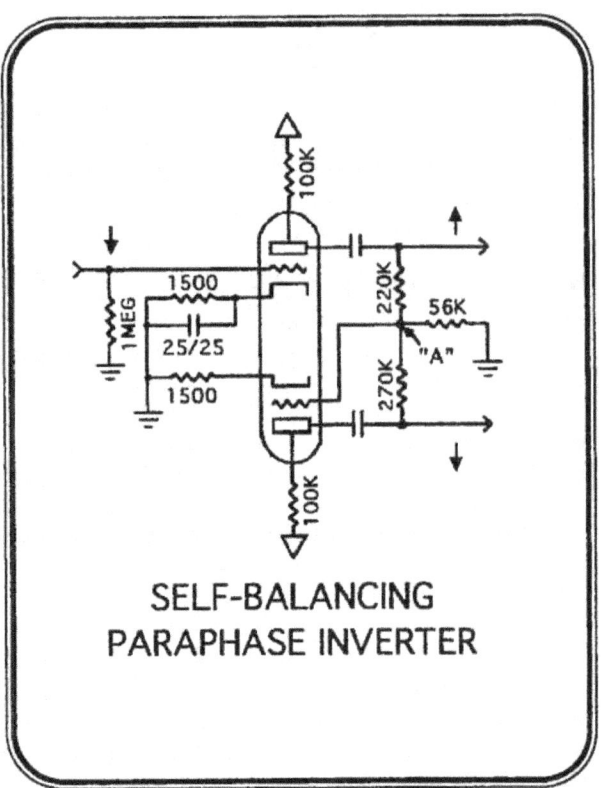

SELF-BALANCING PARAPHASE INVERTER

To understand it, imagine the signals at the power tubes being unequal. If the signal at the top tube is larger than the bottom signal, it will increase the IR Drop through the 56K Common Resistor. This higher voltage adds to the signal going to the Grid of the bottom tube, and produces a larger output signal on the bottom side, "forcing" the circuit to better balance itself.

If the signal at the lower power tube is larger, that signal will also appear at the top of the 56K Common Resistor. It will be amplified and inverted by the lower inverter tube, thereby reducing the signal to the lower power tube, and again, "forcing" the circuit to better balance itself.

This circuit can never be completely in balance. If it were, there would be no difference signal at the Cathode Resistor to feed the Grid of the lower inverter tube, defeating the purpose of the whole circuit.

Fender "solved" this by using 220K to the top tube, and 270K to the bottom, while increasing the drive to the top power tube by using a 25/25 bypass cap on the cathode of top inverter tube. No cap was used on the cathode of the bottom tube, which had a separate Cathode Resistor. This is not convential practice, although this circuit will maintain better balance than the conventional paraphase inverter.

CATHODYNE, OR SPLIT LOAD INVERTER

The next stage of inverter development was found on the 1955 "5E3" Deluxe. This is the Cathodyne, or Split Load Inverter. This inverter is seen on a lot of other Tweed era amps, and on almost all Orange and Hiwatt amps. It's part of the sound of THE two British power houses. It is simplicity itself. Remember, as a signal goes through a tube stage from Grid to Plate, the signal's phase is reversed. When the signal goes from the Grid to the Cathode, the Phase is unreversed. This means we can get a push-pull signal off of one tube. The outputs are opposite, but are they in balance? By dividing the Plate Resistor in half, and putting half of it on the Cathode, and half on the Plate, we create a cross between a Plate Loaded Triode, and a Grounded Plate Triode

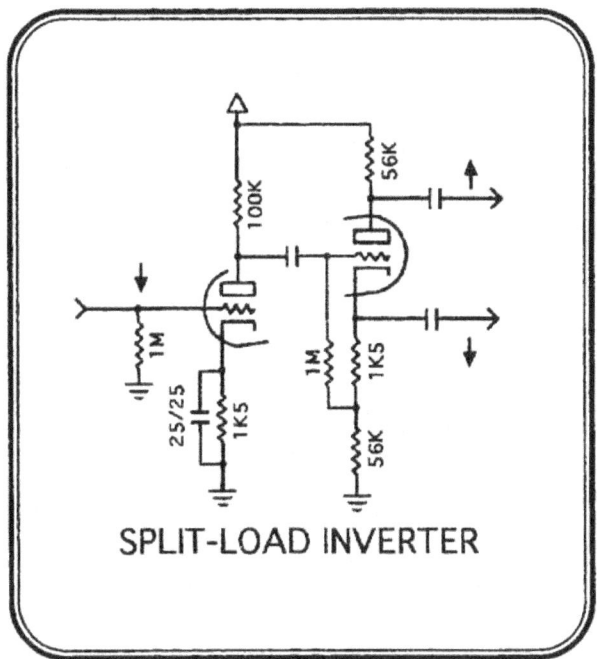

SPLIT-LOAD INVERTER

(Cathode Follower), at the same time. The same current that flows through Cathode Resistor, also flows through the tube's Plate resistance, and the Plate Resistor. Ohm's law says that the voltage across each resistor is equal. Is this the perfect Inverter? Well, nothing's perfect.

The Cathodyne inverter is basically a Cathode Follower. Cathode Followers have gains of less than 1, in other words, a loss of gain. The gain off the plate is no better. You'll remember, that by the time you get a Cathode Resistor up to 47K, you're at Unity Gain.

To compensate for this lack of gain, the cathodyne Inverter needs a Driver in front of it. That's the purpose of the first triode in the circuit. But, if we're going to use two tubes anyway, maybe we can make better use of them. This led to the design of the most complicated, confusing, and popular Inverter in all of electronics, the "Long Tailed Pair."

THE LONG TAILED PAIR

The "Long Tailed Pair" is the most common inverter used. You'll find it in solid state amps and even in opamps. To understand it, let's look at what led to it, the Differential Amplifier.

THE DIFFERENTIAL AMPLIFIER

"Single-Ended" Amplifiers are used with Unbalanced Circuits. You amplify the signal relative to Ground. When you have a Balanced Line, you need something that will amplify one signal realtive to another signal, the foundation of the twisted pair. In other words. you'll need two, complete, symetrical amplifiers. There is studio gear built this way. The drawing of the Differential Amplifier shows such a configuration.

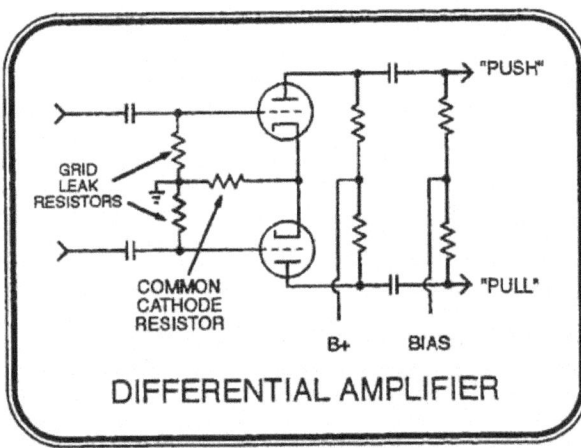

DIFFERENTIAL AMPLIFIER

The circuit has two input signal ports, two Plate Resistors, two Coupling Caps, and two output signal lines marked Push and Pull. What it doesn't have is two Cathode Resistors. Instead, it shares just one. Remember, these engineers were trying to save money, and a clever scheme to avoid using a bypass capacitor on the Common Cathode Resistor, was to share a common one.

Here's how it works. As a signal comes onto the top Grid, it causes Current to flow through the top tube, and subsequently, the Cathode Resistor. As it does this, the DC voltage on the resistor starts to rise as a result of the IR Drop. This makes the Grid look more and more Negative, which is a good thing for biasing purposes, but a bad one from the standpoint that this bias voltage tends to turn off the tube, limiting Gain. It's like its own little compressor.

The way to avoid this problem is to "bypass" the Cathode Resistor with a Capacitor, so the Signal can go to Ground, while the DC component of the Voltage that's useful for biasing is retained.

This isn't necessary in the Differential Amplifier. When a Push-Pull signal is applied to it, one of the Grids moves upward, as the other Grid goes down. Since the Cathodes are tied together, no AC signal appears across the Common Cathode Resistor. It's as if the Cathode Resistor had a Bypass Capacitor across it. Each tube can now develop its maximum gain, without the "compression" effect.

If the SAME signal is applied to both Grids, then the effect is cumulative, and the rise in Common Cathode Voltage offsets the rising Signal Voltages appearing at each Grid. The Signal, by raising the Current through the Common Cathode Resistor, reduces the Grid to Cathode Voltage of both tubes, and reduces their Gains. Gain Cancellation is the reason we use Cathode Caps in the first place.

So the effect is, a balanced, opposing signal, creates a high gain situation in each tube, and an unbalanced, contracting signal, creates a low gain situation in each tube. The result is an increase in dynamics. If the Common Cathode Resistor is made larger, the Gain for the Unbalanced signal drops quickly while the Balanced Gain remains at its maximum.

If one of the Input Grids is Grounded, you have the origins of the "Long Tailed Pair."

GROUNDED GRID
or CATHODE FED

The three ways to hook up a vacuum tube are:

    1. The Grounded-Cathode
       or Plate-Loaded Amplifier.

    2. The Grounded Grid
       or Cathode-Fed Amplifier.

    3. The Grounded Plate
       or Cathode Follower.

If you look at the redrawn circuit, the top tube looks like a Split Load Inverter, with its Cathode Output connected to the Cathode of the bottom tube. This makes the bottom tube a Cathode Fed Amplifier, also known as a Grounded Grid. This circuit has a lot of Gain, but doesn't invert the signal.

As the Cathode Resistor is made larger, circuit balance improves. Funny, the other guys talk about the importance of balancing the Phase Inverter, and never mention this way of doing it. The outputs from this circuit are nearly equal and opposite. To improve the Gain Balance between the stages, Fender made the top Plate Resistor 82K, and the bottom Plate

Resistor 100K. The effect was to slightly lower the Gain of the top tube.

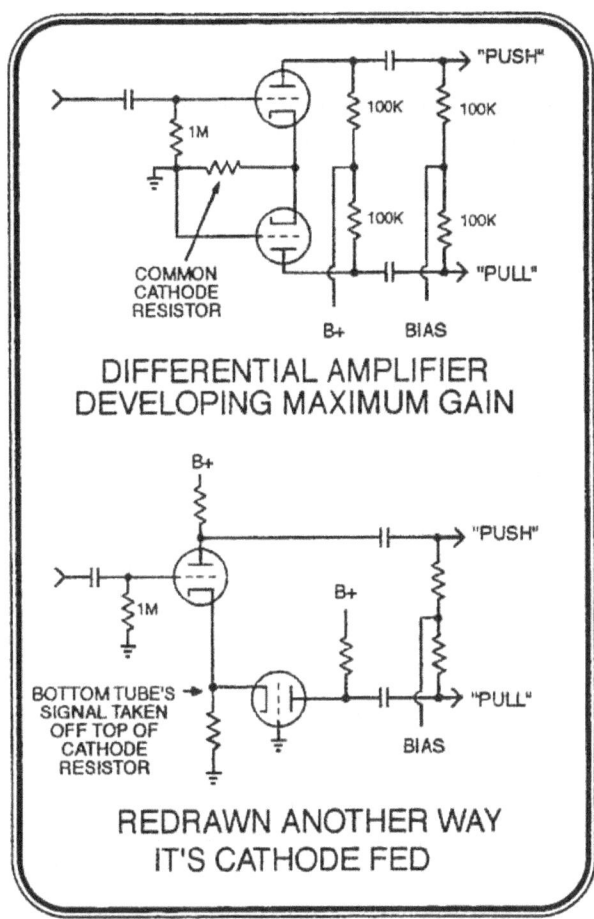

DIFFERENTIAL AMPLIFIER
DEVELOPING MAXIMUM GAIN

REDRAWN ANOTHER WAY
IT'S CATHODE FED

If the Cathode Resistor is made very large, and connected to a large Negative Voltage, this Inverter is called "Long Tailed." It has a long tail reaching down to the Negative Rail? The Long-tailed Inverter is found in almost all amps, Fender or otherwise, built since 1960. The Load (Plate) Resistor and Output Signal are in the Plate circuit; the Signal enters through the Cathode (that's why it's called "Cathode Fed"); and, in a Fender amp, the normally Grounded Grid is used for injecting a Negative Feedback signal. It also permits stable operation at higher frequencies, due to the shielding effect of the Grounded Grid.

CHAPTER 19
FEEDBACK LOOPS

Feedback could be a book in itself, and there are secrets to be found here. This will not be an exhaustive review, but rather some of the highlights.

The concept of Feedback was developed by H.S. Black and his team at Bell Labs in the 1920's. The idea is to return some of the Output Signal to the front of the circuit. It can be either Negative or Positive, and either Voltage or Current, totaling four different types. These four types of Feedback can also be used in combination; Negative Voltage Feedback, with Positive Current Feedback, etc.

If the effect of feedback is to increase the Gain, the Feedback is Positive; if it decreases Gain, it's Negative. The amount of Feedback is annotated in dB and equals the amount of Gain increased with Positive Feedback, or decreased with Negative Feedback. Negative Feedback of 20dB means a reduction in Gain of 20dB.

While the idea of returning some output signal to the front of the amp, and in doing so, lower the Gain of the stage, seems silly, the radio industry quickly adopted it as a way to raise the operating frequency of radio equipment.

NEGATIVE FEEDBACK

A Negative Feedback circuit is sometimes called an Inverse Feedback circuit, or a Degenerative circuit. It's made by taking a portion of a tube's output voltage and applying it to the input of the same tube, or an earlier stage in the circuit, in opposite phase to the original signal. The three important advantages of Feedback are:

1. Reduced Distortion from each stage included within the Feedback circuit.

2. Reduction in the variations in gain due to changes in line voltage, possi ble differences between tubes of the same type, or variations in the values of circuit constants included in the Feedback circuit.

3. The Input and Output Impedance of a circuit with Feedback can be adjust ed, and matched to a Load.

Ignoring Phase effects, the magnitudes of all the harmonics, all the intermodulation products, and all the Hum introduced by the amplifier are reduced by Negative Voltage Feedback, and the Input Impedance increased, in the same proportion that the Gain is reduced. Feedback has no effect on Thermal Noise originating at the Input.

The typical Common Cathode Triode preamp stage has a Voltage Gain of 48, a Frequency Response of 30KHz, an Output Impedance of 2K, and distortion of 2%. The same circuit with Negative Feedback applied, changes to a Voltage Gain of 8.2, a Frequency Response of 100KHz, an Output Impedance of 500, and a Distortion figure of less than 0.4%

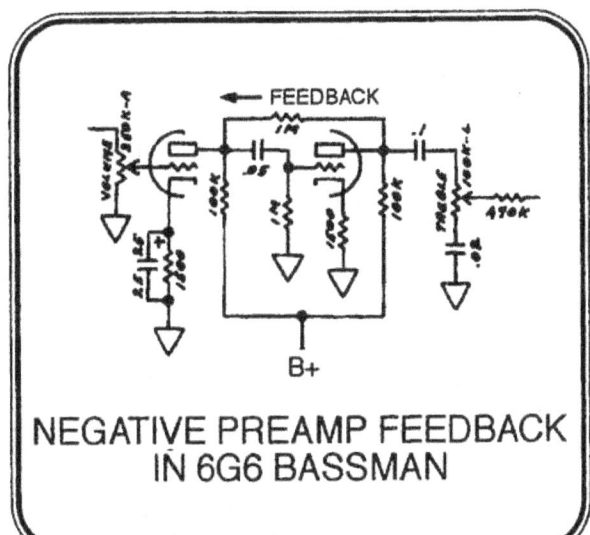

NEGATIVE PREAMP FEEDBACK IN 6G6 BASSMAN

POSITIVE FEEDBACK

Positive feedback is used to convert an amplifying valve into an oscillator, as used in the Vibrato circuit. It's also the foundation of all "Active" tone controls, such as Parametrics (I told you this could be a book in itself). Positive Feedback causes a circuit to resonate and form a "peak" at the Boost/Cut Center Frequency.

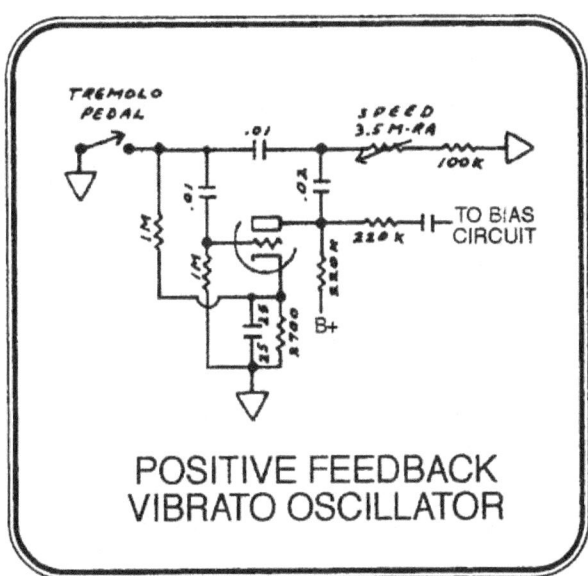

POSITIVE FEEDBACK VIBRATO OSCILLATOR

There are two types of Negative Feedback circuits; the Constant-Voltage, and the Constant-Current type. If the voltage is proportional to the Output Voltage, it's called Voltage Feedback; if it's proportional to the Current through the Load, then it's called Current Feedback. These two are, of course, identical if the Load is a constant resistance, since the Voltage and Current are then proportional.

Negative Feedback is used in audio amplifiers to reduce Distortion in the Output Stage where the Load Impedance on the tube is a Loudspeaker. Since the speaker's Impedance is not constant for all audio frequencies, the Load Impedance on the output tubes varies with frequency. When the output tube is a Pentode or Beam Power Tube having high Plate Resistance, this variation in Plate Load Impedance can produce considerable frequency distortion. Negative Feedback reduces this distortion, and reduces the amps Output Impedance, improving the amp's damping factor (see the section on speakers). It also tends to damp high frequency oscillations.

FEEDBACK CIRCUITS

Suppose that when a Signal is applied to the Input Grid, the resulting Plate Current has an irregularity in its positive half-cycle. This irregularity doesn't match the input signal, and is, therefore, distortion. After passing through the tube, the Plate Voltage's waveform is the inverse of the input Signal. This happens because a Plate Current increase produces an increase in the Voltage Drop across the Plate Resistor (ignoring the next stage's Grid Resistor for the moment). The resulting Output Voltage at the Plate is the difference between the Voltage Drop across the Plate Resistor, and the Supply Voltage; thus, when Plate Current goes up, Plate Voltage goes down; when Plate Current goes down, Plate Voltage goes up.

HOW TUBES REALLY OPERATE, REPRISE

Remember, the Plate Resistor and the Supply Voltage don't change, but the tube's internal Plate ResisTANCE (not Resistor) changes its value. The Plate Resistance (the tube) is at the bottom, with the Plate Resistor on top. If the Plate Resistance goes down, the Plate Voltage is now divided across a the still larger Plate Resistor, and a now smaller Plate Resistance (the tube). Power Supply Current (Plate Current) goes up as the total resistance stack is now smaller. More Current means a larger IR Drop across the Plate Resistor compensated for by a smaller IR Drop across the tube. thus, the IR Drop is divided.

This puts the Plate at a lower Voltage as it tracks the Plate Resistance down, much the same as a Pot divides a Voltage across it as the Wiper is moved. As I started with, remember, the Plate Resistor, and the Supply Voltage don't change. The Power Supply Voltage is divided up between the Plate Resistance, and the Plate Resistor. If you can remember this, you'll know EXACTLY how a tube operates.

BACK TO FEEDBACK

Now add Feedback to the amplifier. The Voltage, after being divided down by the Feedback Resistor, puts a signal back onto the Grid that has the same

Waveform and Phase as the Signal on the Plate (the Output), but is smaller in magnitude. This Voltage applied to the Grid produces a component of Plate Current, that would act to cancel the original irregularity caused by passing through the circuit. This is how Negative Feedback acts to correct any distortion in an output signal.

Consequently, when Negative Feedback is applied to an amplifier there's a decrease in Gain or Power Sensitivity, as well as the decrease in Distortion. This requires a higher driving voltage to be applied to the power tube's Grid, to obtain Full Power Output, but that Output is at a lower Distortion level.

CURRENT FEEDBACK

Cathode-Follower circuits allow the design of High Input Impedance, High Output Voltage, Low Output Impedance circuits, useful in Effects Loops, and other situations where you need a buffer. Due to the Negative Degenerative Current Feedback, inherent in the design, very low distortion can be obtained. This also accounts for their Gain being less than Unity.

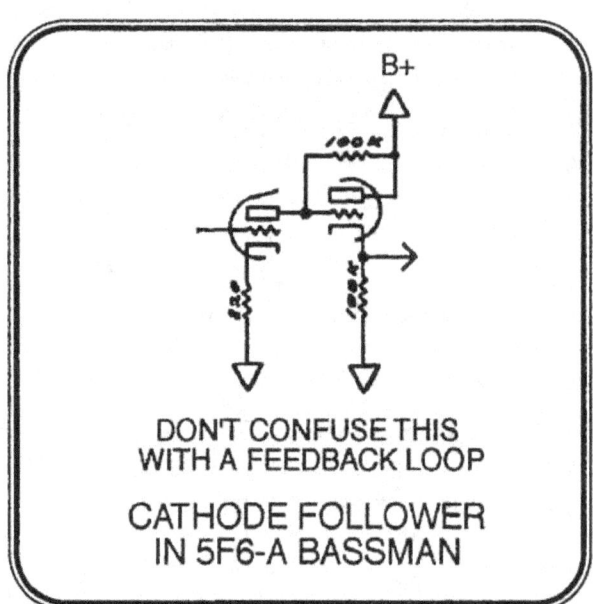

DON'T CONFUSE THIS
WITH A FEEDBACK LOOP

CATHODE FOLLOWER
IN 5F6-A BASSMAN

The "Current" part of the Feedback is actually a "Current Reading" taken off the Cathode Resistor, and converted to a Voltage just like the one ohm Resistors attached to the Cathodes of your Power Tubes to measure the Idle Bias Current. The Cathode Resistor acts as a "shunt," or a resistor that creates a Voltage Drop. The Voltage Drop is directly applied to the tube's Cathode (as explained in Chapters 8-11), and affects the Preamp Tube's Bias, reducing the Gain of the stage; and that's Feedback, isn't it?

Current Feedback can also be obtained by omitting Cathode Bypass Capacitors across a preamp or power tube's Cathode Resistor, reducing distortion by introducing degeneration into the circuit. However, the use of an un-bypassed Resistor decreases Gain and Power Sensitivity, and should not exceed 10% of an amplifiers total Feedback network. It's best to do this in the 1st or 2nd stage, and not on the Power tubes. So, you never knew that by removing a Cathode Cap, you just changed the Feedback network.

NEGATIVE CURRENT FEEDBACK DAMPING CONTROLS

Another way of applying Current Feedback is to place a very small Resistance in Series with the Output Transformer Secondary circuit (the Speaker Side), and letting the Output Current generate a Voltage that can be applied to an earlier stage. These circuits are commonly called "Damping Controls." These circuits decrease Gain and Distortion but increase the Source Impedance (Output Impedance) of the circuit. Consequently, the Output Voltage rises at the Resonant Frequency of the loudspeaker and accentuates hangover effects. In other words, it destroys the Damping Factor, or how tightly the amp can control the movements of the Speaker Cone.

Current feedback is undesirable in transformer loaded amplifiers for another reason. It tends to stabilize the output transformer's magnetizing current (i.e. make it sinusoidal) and thus produce a distorted

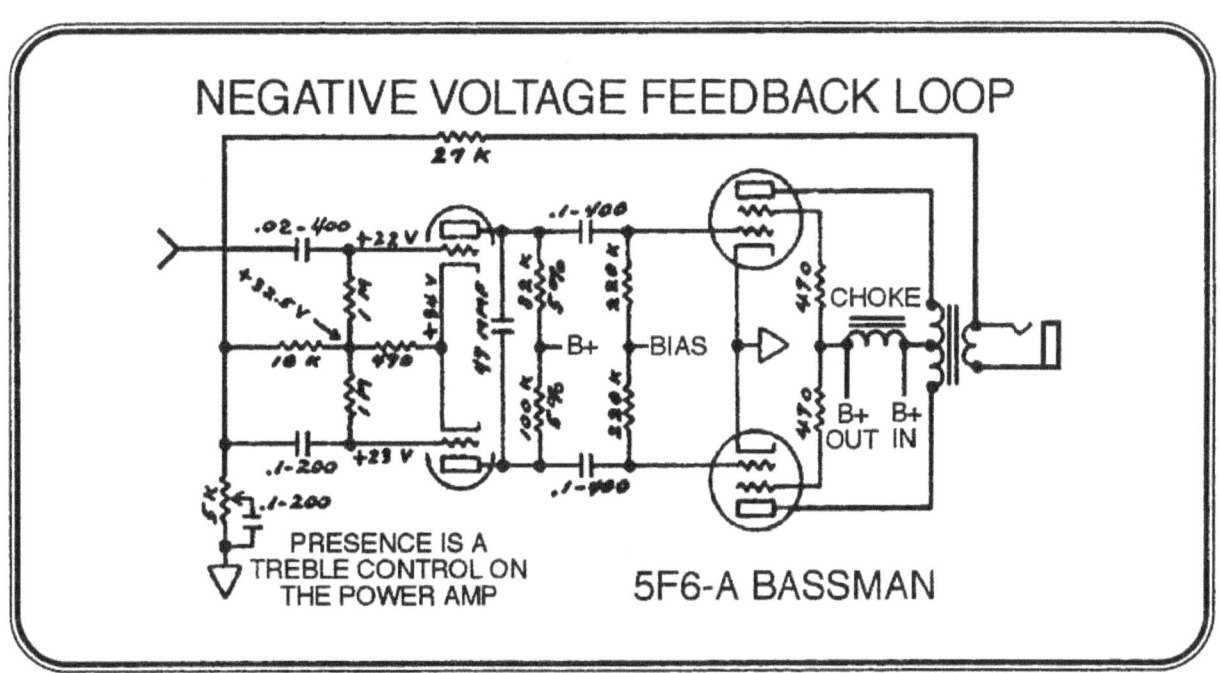

NEGATIVE VOLTAGE FEEDBACK LOOP

PRESENCE IS A
TREBLE CONTROL ON
THE POWER AMP

5F6-A BASSMAN

Output Voltage.

This circuit also increases the tube's Plate Resistance proportionally greater than the decrease in Gain. It should be remembered that the tube itself doesn't change, but the Output Impedance may be adjusted to equal the load resistance. Harmonic Distortion is reduced in the same proportion that the Gain is reduced

COMBINATIONS OF FEEDBACK

Negative Current feedback may be combined with Positive Voltage Feedback to give very high effective Plate Resistance.

Negative Voltage Feedback may be combined with Positive Current Feedback to decrease the Plate Resistance to zero, or even to make it Negative. This is little used in amplifiers because the Positive Current Feedback increases Harmonic Distortion. However, it is possible to combine Negative Feedback in the Output Stage with Positive Feedback in an earlier stage to give useful results.

The Distortion in a two or three stage amplifier is mainly in the Output Stage, and the Distortion in a well designed first stage will be relatively small. It is practical to apply Positive Voltage Feedback to the first stage only, and then to apply Negative Voltage Feedback over two or three stages to get very low Distortion and a low Output Impedance.

BALANCED FEEDBACK AMPLIFIERS

A Balanced Feedback Amplifier is one using both Positive and Negative Feedback in such proportions that the overall Gain with Feedback is equal to that without any Feedback. In other words, the effects of the Negative Feedback are in Balance with the effects of the Positive Feedback, resulting in no Gain change, but the Output Impedance may be adjusted to equal the Load Impedance.

CHAPTER 20
STANDBY SWITCHES

CATHODE STRIPPING

Inside a vacuum tube, a strong positive Voltage on the Plate is used to attract a flow of electrons from the Cathode. When the amplifier is first turned on, the Cathode is cold. It takes 30 to 60 seconds for the heater (filament) to heat the Cathode to a point (1050⁻ Kelvin) where a charge of electrons (a space charge) is created around the Cathode. This space charge supplies the electrons that flow to the Plate. With the space charge properly established, the Cathode will actually renew itself. If the Cathode is cold, the high Voltage on the Plate will steal electrons from the cathode's metal, causing the cathode's oxide coating to erode away. This "stripping" of electrons rapidly ages the tube. With the price of tubes today, every effort should be made to preserve their life. Of course preserving your tubes reduces the demand for new ones, increasing their cost, and decreasing their availability, but that's how capitalism works.

DISCHARGING THE AMP

Another less known use of the standby switch is to discharge the power supply caps when you turn the amp off. Since the standby switch interrupts the flow of electrons from the power supply storage caps to the tube's Plates, any energy stored in the caps will stay there for awhile. If you turn the AC power switch off first, you'll notice the sound will slowly fade away. This is the electrical energy stored in the caps being used up. This is a great safety measure to insure that if you do open the chassis, there won't be any high Voltage stored up waiting to knock you down. The energy stored in the capacitor can sit there for many days.

WARNING: This only works if the tubes are hot, AND if you have a high tension fuse, it can't be blown. If you turn an amp on, and immediately turn it off, the power tubes weren't given sufficient time to heat up to operating temperatures, where they'll be able to conduct away the stored energy. If your high tension fuse is open, it won't allow the energy stored to drain off. Never work on an amplifier if you don't know exactly what you're doing. If you think 120 Volts AC from the wall socket hurts, try 600 Volts DC. AC might throw you away, breaking contact, but DC holds on to you until you're good and dead.

I personally work on a wooden bench, with the amp on a grounded conductive rubber mat wired to the fuse box's ground lug, in rubber shoes, on a rubber mat, with a ground fault interrupter, a fused variac, through a fused Ampere meter, and I keep one hand in my pocket. The reason you keep one hand in your pocket is so that you don't accidentally grap something that will ground you, completing the circuit and causing Current to flow through your heart.

SAFETY

When working on an amp, always check all of the power supply caps using a Volt meter to verify that they are sufficiently discharged before continuing. Never discharge caps by shorting them out, unless you know that most of the energy stored in them is already gone, or you're about to throw them in the garbage. Shorting out caps can seriously damage them through overheating, just like shorting out a battery can start a fire. I once jumped a car with a shorted starting motor. After the two batteries were connected, the dead car was cranked up and the jumper cables promptly arc welded themselves together. The military has rules about mixing fresh batteries with used batteries, or different types of batteries in flashlights. If the fresh battery rapidly discharges itself through a weak battery, the flashlight can explode, and/or burn. Very embarrassing when piloting your plane at forty thousand feet! My point is that capacitors are not designed for the heat generated when they are shorted out.

CAPACITORS

While on the subject of capacitors I'd like to say that a large electrolytic capacitor is better than a small electrolytic capacitor of the same value. The large cap contains more electrolyte which boils off when overheated. With a larger reservoir, the large caps last longer. It also has more surface area to dissipate any heat generated, staying cool, longer. Remember this when you're putting small imported caps into your amp. I only use American made Sprague caps.

RESIDUAL CHARGE

If you follow my suggestion about letting the amp discharge itself through the standby switch, you'll notice that about five to twenty Volts likes to stay on the electrolytic caps. Then, if you short the caps out with a jumper wire, you'll find zero Volts on the caps. After the jumper is removed, the Voltage starts to rise again! What is it? It's called a residual charge. What happens is that some electrons bury themselves into the insulating layers that lie between the conducting layers of the capacitor. These electrons slowly come back out of the insulator and recharge the capacitor's Plates. This is a normal phenomenon, and the small Voltage won't hurt you.

FORMING CAPACITORS

Electrolytic capacitors go through a process called forming, where they increase their capacity, and reduce their leakage, after they become charged. The hype is that caps must be properly formed or they'll never sound good. The truth is that Fender never formed his caps, and once a cap is formed, it doesn't stay formed anyway. If you don't use the amp for a couple of weeks, the caps will need to be reformed again. They will form themselves automatically the next time you turn the amp on.

If there is any difference in the sound of an amp with formed or unformed capacitors, it's because formed capacitors have slightly greater capacity, and slightly less DC leakage.

One nice thing about amps with the GZ34/5AR4 type tube rectifiers is that the large DC Voltages to the caps, and tubes, come up slowly as the rectifier heats up. This automatically gives the caps, and the tubes, a soft start every time you turn the amp on, and tends to negate what I said about standby switches saving the power tubes. You should also already know that a tube amp will sound better after it's been warmed up for 30 minutes.

So to summarize, the proper way to turn an amp on is power switch first; Wait 30 to 60 seconds, then turn on the standby switch. When turning an amp off, turn the power switch off first, wait 5 to 10 seconds, and then turn the standby switch off, so it's in the proper "off" position the next time you use the amp. Get in the habit of always checking the position of the switches before you turn your amp on. If you accidentally turn the power on, and then notice the standby switch is already on, turn the standby switch off as soon as possible. It will help save your tubes. In short, the power switch should always be first on, and first off.

CHAPTER 21
GROUND LOOPS, NOISE & SAFETY

In this chapter, we're going to review electrical wiring and safety principles. The reason for this is there's a lot of misinformation out there, and this is a critical safety factor that needs to be addressed.

AC POWER

Electricity is delivered to your house as a Black Hot wire carrying 120 Volts of Alternating Current. The White Neutral wire derives from a Ground Connection at the Service Entrance (Fuse Box). All the Current that flows through the Black wire, alternately flows through the White wire, being pulled up out of the Ground. This allows the Electric Utility to string only one wire to your house, instead of two.

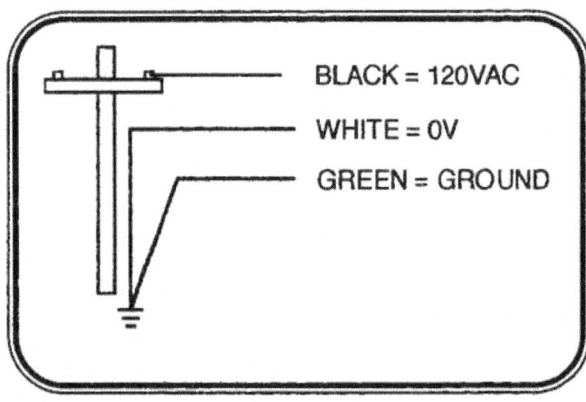

BLACK = 120VAC
WHITE = 0V
GREEN = GROUND

Most houses have a two wire service, each one a separate 120 Volt circuit, with their phases being 180 degrees apart. If you use both "Hot Legs," you get a 240 Volt circuit, with Ground sitting between the two phases.

If you have "Three Phase," you have three 120 Volt circuits, each 120 degrees apart, producing 208 Volts.

The thing to remember is, with a 120 Volt circuit, the White Neutral wire is very similar to the Green Ground wire. You SHOULD be able to grab it and not get a shock. It's the Black one you want to watch out for. This all assumes that the wiring is correct. Every musician should carry one of those plug in wiring testers. We worked a job at a fancy hotel, and found 120 Volts on the Ground pin! Test before plugging in.

AC PLUGS

When I was growing up, electrical plugs had two prongs of the same size. Some heavy equipment had three prongs, but in general, products used around the house only had two. This included guitar amplifiers. Both prongs were the same size, and it didn't matter which way they were plugged in.

Of course, if you plugged it in the wrong way, the amp would be noisier. The Ground Switch solved this by attaching a .047/600v (it's AC rating is lower) to one side of the power line; or the other. This forms an AC path from the Chassis to the AC line. Noise reduction through grounding, without actually connecting a Ground wire to the Chassis.

If two amps are being used, as with a guitar amp and a PA head, the polarities of the Ground Switches have to be the same. If they're not, an AC leakage path forms between your fingers, which are touching the amplifier chassis through the "grounded" guitar strings, and your lips, touching the PA amplifier chassis through the shell of the microphone. I guess we've all been there - done that.

POLARIZED PLUGS

Some plugs come with one prong wider than the other. These only plug into the outlet one way. They're called "Polarized" Plugs. They assure that the Neutral White wire (the larger prong) is correctly connected to the appliance being plugged in. This polarity assures that the "chassis" of the device is not connected to the Black wire, preventing a potential electrocution scenario when you touch the metal chassis and something innocent, like a water faucet. You become the conductor to Ground, and if you don't draw enough current to blow the fuse, you might stay connected forever. You'll see these plugs on TV's, and portable power tools.

DOUBLE INSULATED

A safety scheme came along called "Double-Insulated." A portable drill might have no exposed metal at all. In these cases, all the electrically conductive metal parts are concealed under a plastic body. It has a two pronged plug with no polarity. It increases the safety factor for devices that might be used outdoors on wet ground, or on a wet concrete floor.

THREE PRONGS

A Three Prong Plug is the only way to go for me. It has a Black Hot wire, a White Neutral wire, and a separate Green

Ground wire. The White and Black wires attach to the amp's Power Transformer through the Fuse, and the Power Switch. The Green wire is connected directly to the metal chassis of the amp. If anything goes wrong with the amp, the chassis is firmly attached to Ground, forming a short circuit path for the "error" currents, and assuring that the current will flow to Ground through the Green wire, and not through you. The Green wire will probably always have a lower Resistance to Ground than your body does.

VINTAGE CONVERSION

Converting a vintage Fender amp to a Three Pronged Power Cord requires more than just soldering the new Black and White wires to the Fuse & Power Switch, and adding the Green wire to the Chassis. When that amp was built, it had an unpolarized, two prong plug. Since the designer never knew which way someone might plug the amp in, it didn't matter which side the Fuse or the AC Power Switch were put in. Fender put the AC Switch on one side, and the Fuse on the other.

This is incorrect if you're using a Polarized, or a Three Prong plug. The Fuse and the AC Switch should both be on the Black wire, with the Fuse coming first. A Switch or a Fuse should never be put in the White Neutral wire circuit. Anyone who wires houses knows this.

If the Fuse is put in the Neutral path, and the Fuse blows, you might think the amp's dead. A very logical conclusion, since the Fuse is blown. But the Fuse disconnected the Neutral circuit, not the Hot circuit.

Now suppose there's a short to metal

inside one of the Transformers? A likely occurrence with those Paper Wound ones. Now you have AC Voltage on the amplifier's Chassis; the Fuse is blown; and you touch the Chassis. In this case, the Green wire saves you. The Chassis is Grounded through the Third Prong of the AC Plug. Any Voltage on the Chassis is shorted out to Ground, and you're saved!

BLACK
SWITCH
FUSE
WHITE

OLD STYLE

BLACK
SWITCH
GREEN
FUSE
WHITE

INCORRECT CONVERSION

BLACK
FUSE SWITCH
GREEN
WHITE

CORRECT CONVERSION

Except, you thought you'd solve that Ground Loop problem by disconnecting the Ground wire with that handy "Grounding Adapter" you always carry with you.

So, now there's no Green wire protecting you by shorting the chassis to Ground. You walk up to the amp, and touch something metal, like the On/Off Switch. Now, YOU become the path to Ground for all the shorted current. And as an added bonus, YOU HAVE NO FUSE IN THE CIRCUIT! Don't worry. If you draw 15 amps or more, you'll likely blow the Mains Fuse back at the Fuse Box.

If only you hadn't disconnected the Green wire with that "Floating Adapter." Any Current flowing from the Transformer to the Chassis, would have been shorted to Ground by the Green Ground wire, blowing the Fuse back at the Fuse Box. Remember, the smaller amplifier Fuse, which was left in the Neutral Circuit, is already blown. The problem is, it isn't where it needs to be; in the Hot Circuit. Now, you might see why we need to make more changes to this Vintage Conversion.

FLOATING GROUND ADAPTERS

These "Adapters" really bug me. They were designed to allow you to screw the little tab to the outlet screw on a Two Prong Outlet, forming a Ground connection for a Three Prong Plug. It's hard to make something foolproof, because fools are so ingenious!

If you have a Ground Loop, solve the Ground Loop problem. Thinking you're doing that by "Floating" the Ground is just ignorant. Ground Loops are not complicated if you understand the problem. Finding the cause can be easy, using a little logic.

THERE"S ALWAYS
AN EXCEPTION

In recording studios, the amps are ungrounded on purpose. A Chassis Ground wire is run from the amp chassis directly to the Recording Console. This Grounds the amplifier, and eliminates Ground Looping between the AC Grounds, and the Signal Ground going to the Console.

This is why the Thunderfunk Amp comes with a Ground Switch, even though it comes with a Three Prong Plug. Sometimes the Chassis Ground is defeated ON PURPOSE BY PROFESSIONALS WHO KNOW WHAT THEY'RE DOING. Then the Ground Switch can work to reduce noise.

Ground Switches are sometimes called Death Switches, and are not legal in Europe. If the Chassis is actually grounded by using a Three Prong Outlet and a Three Prong Plug, the Ground Switch no longer works. So, it's redundant to Ground the amp, and then worry about which way the Ground Switch is thrown. The only reason to put a Ground Switch on a Grounded amp is to allow you to defeat the Ground, and still have noise reduction. I suppose disconnecting the Ground is why they're called Death Switches.

GROUNDING

I designed an oscilloscope for automotive testing, and it had a "Ring" in it, everytime a spark plug fired. The cause was a Zener Diode that was Grounded one inch away from the scope's Chassis Ground point. It was the last thing attached to the Ground circuit, before the circuit board attached to the Chassis Ground Strap.

Everytime a sparkplug fired, the burst of current through the Ground Return Path caused an "IR Drop" (Current times Resistance = Voltage Drop) through that one inch of copper. This caused the Zener's Ground Reference point to jump. And, since everything was referenced to the Zener, everything bounced along with it. Moving the Zener's Ground point one inch, solved the problem.

Sometimes, when you're hunting noise, especially hum, you should attach a scope probe to some of the "Ground" points. An example would be the Ground lug of an Input Jack. If corrosion has caused a bad connection to the Chassis, a Resistance could be formed, causing the Input Signal Ground to float and pick up some of the surrounding hum. This is readily seen on the scope display. Without a steady Ground reference, you'll get noise superimposed on any signal. This is especially important when you find the noise on the input of the first Gain Stage. Where else could it come from?

GROUND LOOPS

A Ground Loop is formed when the Return Current has a choice of two, or more, Return Paths.

An example I'll use is a two amp setup. Plug each amp into a different wall socket. Now, split your guitar cord into a "Y" and plug into both amps. The noise you hear is a Ground Loop. The Signal Cord has a Ground connection between both amps, and so do the AC Plugs.

The way to eliminate the Ground Loop is to either plug both amps into the same outlet, which should work, or plug one amp into the wall, and plug the second amp into the first. Now, the Grounds all flow "down" to the wall.

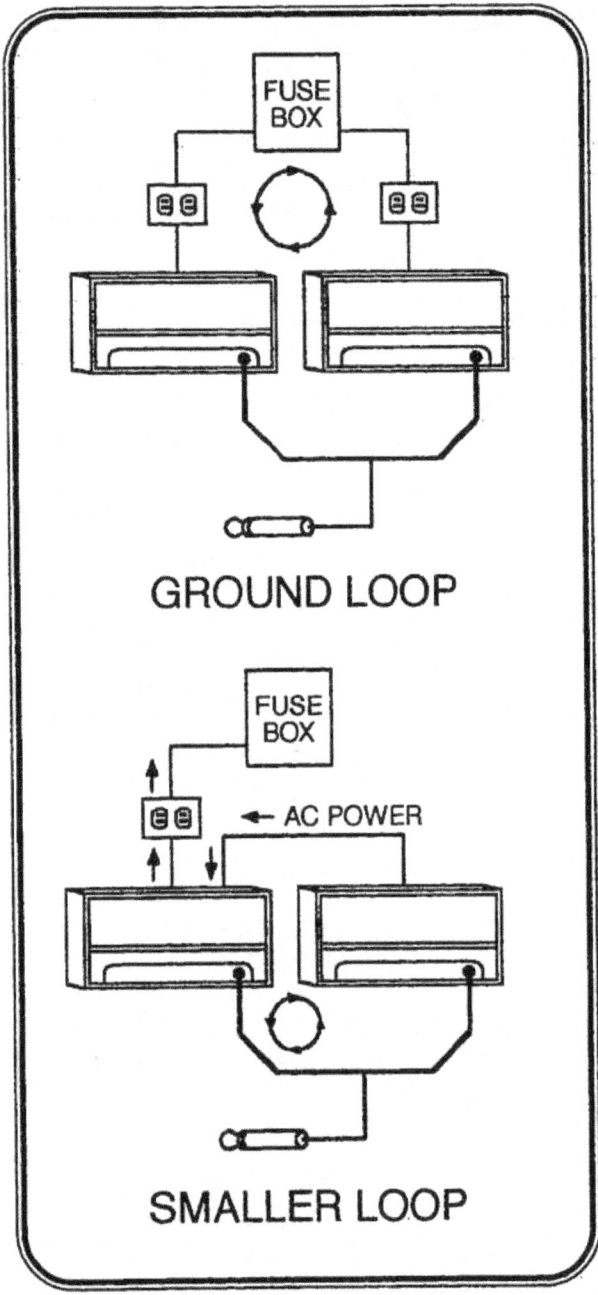

GROUND LOOP

SMALLER LOOP

You can see in the diagram, there's still a "Ground Loop" between the two amp's AC connections, and the "Y" cord. In most cases, this will not cause hum, but it

might, depending on Wire Length, Wire Resistance, and Ground Current flow.

Another example of a potential Ground Loop is a Wireless Receiver, or an Effects unit, and your amp. They're connected by the Signal Cord, and their AC Grounds. Care must be taken in Grounding these units as well.

AC POWER

The power from the wall comes in on the Black wire. After flowing through an amplifier's power transformer, it returns through the White wire to the fuse box. At the Fuse Box, the White wire is attached to the "ground" outside your house. From there the power flows through the ground back to the power generating station. This means the power company only has to run one wire to your house (the Black one), instead of two, and actually uses the ground to provide the Return Path. Inside the amp, the White wire provides the Return Path until it reaches the Fuse Box. In order to generate electricity, electrons are pulled out of the ground, and are balanced by electrons being returned from the customers. (Of course, a Positive voltage is actually a shortage of electrons, which means the power company is creating positive "holes" at the power station, and pulling electrons out of the ground at your house, to satisfy the shortage.)

Two paths are needed, the Black wire and the Ground, because voltage is developed BETWEEN two points; A Hot Black wire, and a Neutral White wire. The amount of power flowing through the Black wire is balanced by the amount of power flowing through the White wire. This balance will become important later on.

GROUND LOOP NOISE

The fact that there is a "return" current is something that is often overlooked. If there were no return current, there would be no Ground Loops. The return current flowing through any grounding system causes "IR" Drops (Current * Resistance = Voltage Drop) that can allow noise to enter the system across the ground return resistance. Examples of this are having multiple pieces of equipment in a rack where the chassis are grounded to the rack rails, and through their respective power cords, or using a "Y" cord to connect a guitar to two amplifiers.

A voltage drop is created between the ground points by the current flowing through and between them. This references the circuits to different ground voltages. Since the signal stages in an amp are referenced to Ground, any signal that appears ON Ground, is added to the output signal. What you get is Ground Loop Noise. It can be 60 cycle hum, or harmonics of 60 cycles which is ground buzz.

It may sound strange to say Ground has a voltage other than zero, since by definition, Ground IS zero. But that's only in theory. Ground is SUPPOSE to be zero volts. But since voltages are read from the test location to Ground, how do you measure the voltage OF Ground? Measure from Ground to Ground? You'll always get zero volts. Both test points may be bouncing all over the place, but they'll both be in sync and will develop no voltage between them.

A design can have more than one Ground. Some designs have Signal Grounds, Digital Grounds, Power Grounds, Chassis Ground, and plain old Ground Grounds, all at the same time. The reason was to separate current from one part of the circuit upsetting the Ground of another.

CHASSIS GROUND

Chassis Ground is the big one, and is usually the "Master" Ground. It is always drawn as shown in Figure 1. In fact, if you see this symbol in a schematic, it is pronounced "Chassis Ground." Measuring the level of another Ground in the circuit would require the negative lead of the voltmeter to be connected to the Chassis Ground point, with the positive lead going to the Ground being tested.

Whenever you connect a voltmeter to an amp under test, you should try and connect the negative meter lead to the point on the chassis where the Green Ground wire from the AC power cord is connected. This is "Chassis Ground." This is the lowest level of Ground, and is what all other Grounds are USUALLY tied too. From here Ground leaves the device and is external to it, and is usually connected directly to the power wiring Ground.

Some designs have a separate Signal Ground that doesn't have anything connected to it other than other Signal Grounds. This Signal Ground is then usually connected to Chassis Ground at only one point. This assures that the Signal reference Grounds are not affected by heavy current flows that might exist in the power circuits.

heavy current flows that might exist in the power circuits of a design.

An example of this might be a logic system using sensors to control heavy electrical motors. The sensors might have a Signal Ground circuit; the logic system a Digital Ground circuit; and the motors, their own "Power" Ground circuit. All these Grounds would be connected together at only one point, usually the Chassis Ground.

If you look at a 19" rack of power equipment, you might see some braided wire straps. These are called "Ground Straps." Their purpose is to provide an extremely low resistance ground path for all the ground currents to flow through, creating as small of an IR Drop as possible.

I was visiting a friend's farm while he was trying to start his tractor. The battery had 12 volts in it, but the starter motor cranked very slowly. Grabbing the battery leads I found that one of them was hot. The corroded wire formed a Resistor that was absorbing the power of the battery before it got to the starter. If he had measured the Voltage at the starter while cranking

he would have found it to be about 8 volts. The large Current flow created a large Voltage Drop through the wire.

When doing instrumentation, a Voltmeter has its own leads called "Sense Leads." These leads would attach to the starter motor and allow you to read the Voltage there, instead of at the battery. This removes the battery cables from the Voltage reading. This is an example of removing high Current paths from instrumentation paths. The same thing applies to Filament leads, which are high Current.

FILAMENT HUM

The tube filaments in a Fender Champ amp are connected by using one "hot" wire connected to one side of the 6.3 volt transformer wires. The other transformer wire is connected to Chassis Ground, along with all the second tube filament pins, forming the filament return path through the Chassis.
Filaments can use a lot of Current, and this Current flow can cause a rather large IR Drop through the amp's chassis grounding system. Removing this current flow by lifting the filament pins from the

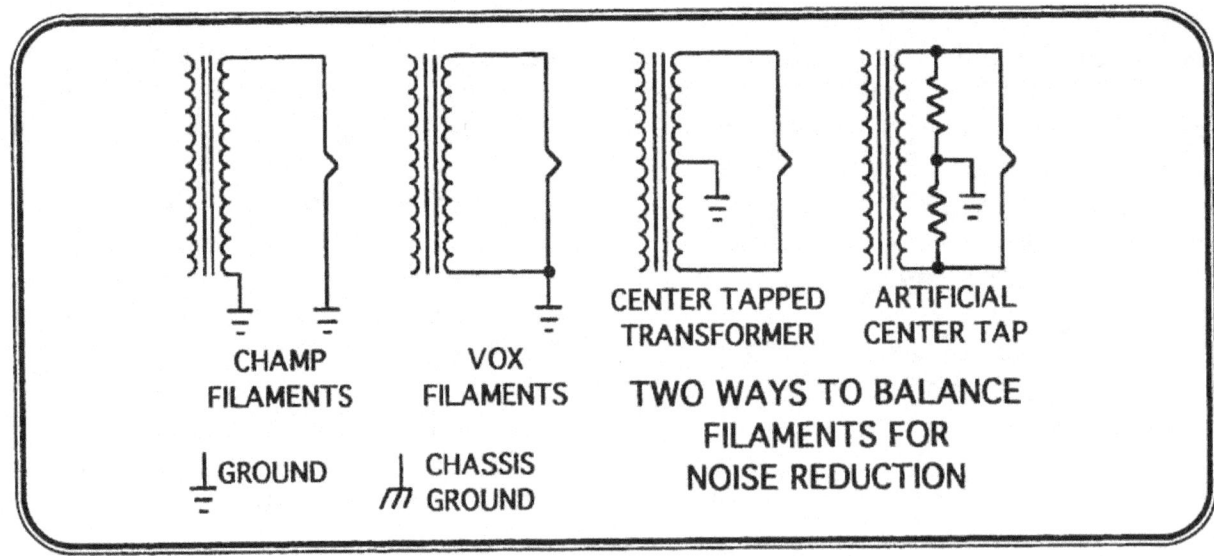

CHAMP FILAMENTS

VOX FILAMENTS

CENTER TAPPED TRANSFORMER

ARTIFICIAL CENTER TAP

TWO WAYS TO BALANCE FILAMENTS FOR NOISE REDUCTION

GROUND

CHASSIS GROUND

Chassis and adding a second filament wire, will reduce the amount of hum in a Champ. Don't forget to reference the filaments to Ground by adding a 100 ohm resistor from each of the filament leads to Ground, or by attaching the Filament Center Tap wire (usually Green/Yellow) to Ground.

The Vox AC-30 uses two wires to connect the filaments, but then one of them is connected to the Chassis (near the fourth preamp tube) creating the same problem as with the Champ. The fix is to disconnect the filaments from Ground, and re-reference them with 100 ohm resistors.

In some cases the filaments need to be referenced to something other than Ground. This is a special case that's rarely seen in instrument amplifiers. The center connection of the two 100 ohm resistors would be connected to a positive or negative DC voltage, much the same as the Bias Voltage is applied to the Power Tube Grids. In fact, Bias Voltage means "offset."

GROUND FAULT INTERRUPTERS

As explained earlier, the amount of Current flowing through the Black wire is suppose to equal the amount of Current flowing through the White wire. If something untoward happens, like you're electrocuting yourself, some of the Current flow is going from the Black wire, through your body, to Ground. This is a "short circuit." You may not draw enough Current to blow the Fuse, before you're killed. So, what to do?

A Ground Fault Interrupter (GFI) measures the balance of Current between the Black and the White wires. Whenever an unbalance occurs, which means that some of the Current from the Black wire is returning to Ground by a path OTHER than the White wire, a Ground Fault Interrupter senses the "fault" and trips its own Circuit Breaker.

I use one on my workbench, and sometimes I'm surprised when it blows. But it's always because of some mistake I've made. It's an extra stage of protection I like to have around. However, I don't rely on the GFI to save me from being careless. A GFI does not sense when you're electrocuting yourself by touching the DC Voltages in the amp. After all, the AC is still in balance. These devices were developed to sense radios falling into bathtubs, and electric hedge trimmers from shorting out when you're working outside on a wet lawn. This is why they're required by Building Code near all sinks, water supplies, and outdoor outlets. I'm not sure if they're required in garages, but wet concrete is an invitation to trouble.

GROUNDING SYSTEMS

There are three main grounding systems; Star, Buss, and Plane.

The first scheme is the Ground Buss. I think you've seen "Buss" fuses. A Ground Buss is a track, bar, or wire, to which all components are attached. The Ground Current flows down the Buss to the master Ground point. A common expression is "Buss Bar." This describes the stiff bar of grounding wire that might run through a circuit.

A Star Ground connects each Ground point to the master Ground point by its own conductor. This scheme is fairly fool proof, but is usually unnecessary in a guitar amp.

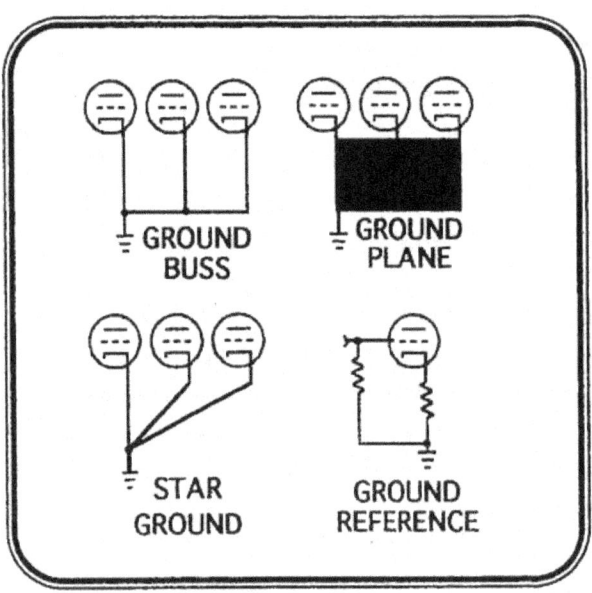

One of the ways to reduce hum is to make sure the Grid Resistor, and the Cathode Resistor are Grounded to the same point. The brass plate behind the pots in a Fender amp is there to provide a low resistance "Buss" Ground for the pots and input jacks. When I modify a Fender amp, I sometimes need to add a Cathode Resistor to a preamp tube. The easiest way to do this is to replace one of the screws that mounts the preamp socket with a 4-40 x 3/8" screw and nylon lock nut, inserting a Ground Lug in the process. Now, attach the Cathode Resistor from the tube socket, to the Ground Lug. Although this places the Cathode Resistor at a distance from the front mounted brass strip, where the input jack, and Grid Resistor are Grounded, there are no noise problems created. You might think of the Chassis as a Ground Plane.

GROUND PLANES

Ground Planes come from circuit board design and refer to the use a an entire layer of circuit board copper to provide Ground. A four layer circuit board might use an entire layer for Ground, and another entire layer for Power. So, you can have Power Planes, as well as Ground Planes. When you connect up a circuit board design, the Power and Ground can become a problem, since they connect to every IC on the board, and you might have +15V, -15V, +5V, +12V, and +48V, as well as multiple Grounds in the same design.

Military avionics might be built on a 12 layer circuit board, with each voltage having its own dedicated layer, making it easy to apply Power and Ground wherever needed. This also opens up the other layers to be devoted to signal lines. The two outside layers have no connections on them at all, keeping all signal and power connections internal. This protects them from damage, as well as snooping.

The way the planes are constructed is to place an oversized pad of copper at every component lead location. This layer is then photographically reversed and combined with a normal pad layer. This provides a component pad, with an annular ring around it, isolating it from the copper plane. At every location that needs to be attached to the Plane, conductors are added from the pad to the Plane. This is then called a "Thermal."

The reason for this is when boards are mass produced, they're "Wave Soldered."

THERMAL PAD
INSIDE
GROUND PLANE

The parts are inserted, and then the boards are passed horizontally over a vat of melted solder. Waves are created in the vat causing the solder to rise up and touch the bottom of the board. If the Plane connections were not "Thermalled" the Plane would absorb the heat, and there would not be enough heat to solder the connections. This would leave "Cold Solder" joints all over the board.

AC GROUNDING MOD

Next, I'll show you how to rewire the most common Fender amps, the Black and Silver faced ones of the '60's, and '70's. It is assumed that any mods made will be done by qualified service personnel.

The drawing shows the typical wiring scheme for a Black or Silver face Fender amp. The illustrations are as viewed when looking into the amp from the back. I normally service this type of amp by placing it on the workbench with the input jacks away from me, to the right, with the power tubes close to me on the left. If you peer into the chassis, this is what you'll see. In other words, the drawing is "upside down."

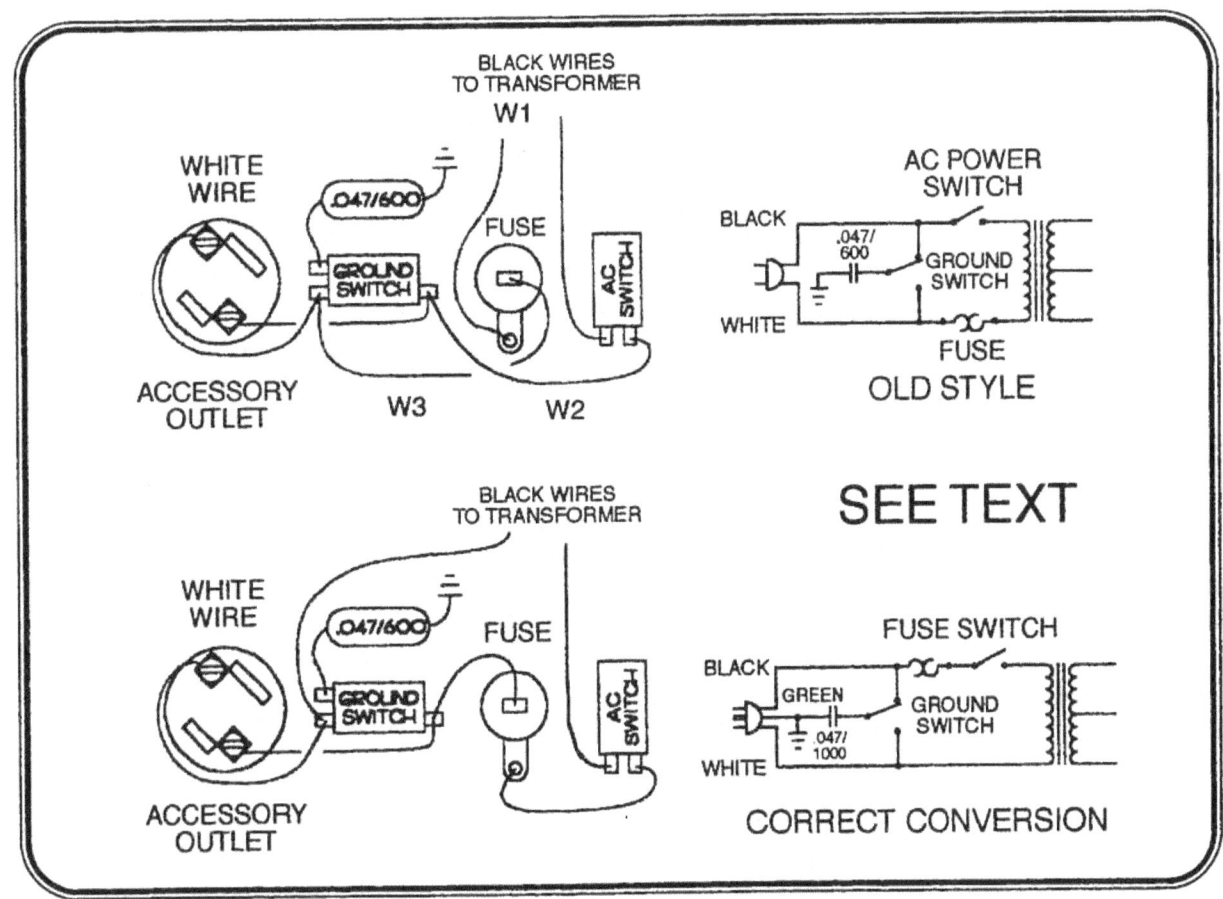

155

Looking at the diagram, this mod is accomplished by going through the following steps:

1. Disconnect the Black transformer wire (sometimes it's White), that goes to the side conductor of the Fuseholder (labeled W1). Leave it disconnected for now.

2. Remove the wire (labeled W2) that goes from the AC Switch to the "Black" side of the Ground Switch, and attach it to the side conductor of the Fuseholder.

3. Remove the end of the wire that goes from the center pin of the Fuseholder to the "White" side of the Ground Switch (labeled W3). Shorten, and attach it to the "Black" side of the Ground Switch.

4. Attach the transformer wire "W1" to the "White" side of the Ground Switch.

5. You're done.

"HOT" FUSEHOLDERS

Remember, the center pin of the fuseholder is the "hot" pin. It connects to the fuseholder's connector that's as far away from your fingers as possible. If you remove the fuse, the hot "Black" wire is still connected to the "back" of the fuseholder. When the fuse is removed, it disconnects from this back pin. Think of it this way; If the fuse being removed isn't blown, and it touches one of the fuseholder's connectors, the front cap of the fuse would be also become "alive." If you connect the Black wire to the side pin of the fusehold-er, the one that connects to the front conductor of the fuseholder, the back cap of the fuse might touch it as you withdraw the fuse. This would then conduct through the unblown fuse, and cause the top of the fuse to become alive - the one that you're holding onto. By connecting the Black wire to the center, or back conductor, the fuse gets disconnected as soon as it's withdrawn. There is a reason to wire the amp correctly.

Also, the "Correct Conversion" in the diagram, doesn't show a Grounded Accessory Outlet. If you're installing a Grounded Power Cord, and are also changing the Accessory Outlet to a Grounded one, the Green wire from the Power Cord would attach to the Green (or Ground) Screw on the outlet, and also to amp's chassis, usually to a solder lug on one of the Power Transformer's mounting screws.

Last month, I showed you how to rewire your Fender's AC supply. In the schematic of the "correct version," the Ground Switch comes before the Fuse. The schematic was drawn this way because the Fender amp is wired this way. The AC Power Cord comes into the amp and immediately goes to the Accessory AC Outlet. From there, it goes to the Ground Switch, and then on to the Fuse and Power Switch. Moving the Fuse and Switch to the Black side of the power line is accomplished simply, with a few changes. However, this leaves the Ground Switch before the Fuse.

To really do it right, you should remove the Ground Switch, or remove the .047

cap that connects the Ground Switch to

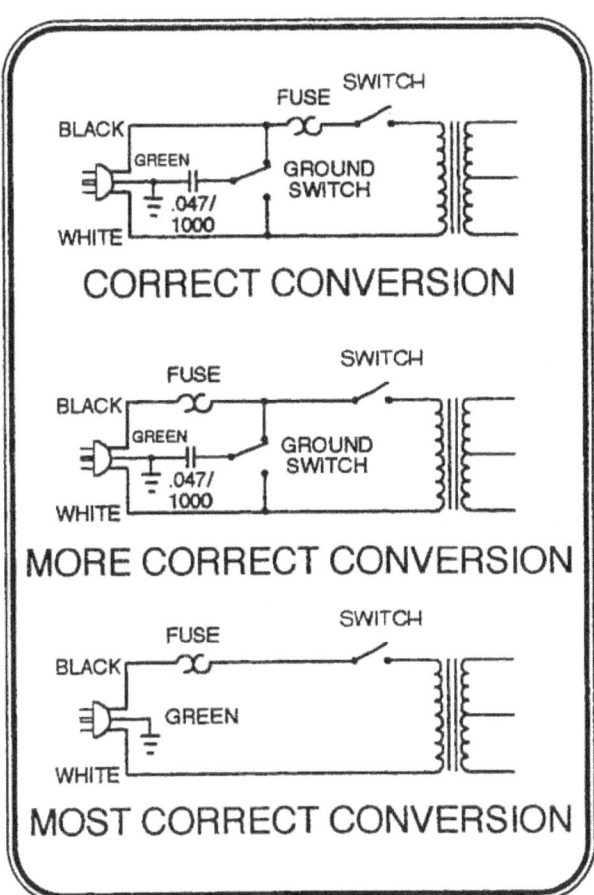

CORRECT CONVERSION

MORE CORRECT CONVERSION

MOST CORRECT CONVERSION

the Chassis. This is the part, that if it failed, could put AC onto the chassis, without the protection of the amp's Fuse. That's the reason for putting the Ground Switch AFTER the Fuse. Do it either way you want. With the addition of a Grounded Power Cord, you're way ahead how the amp was wired originally. So, to be absolutely correct, you should follow the second schematic - or the third one.

CONCLUSION

Grounding is a subject in itself, and I've touched on a few examples of problem solving. If these examples don't work, there's other things you might try. The thing is, a Ground Plane is a very good grounding system, and it's nothing more than a large number of Ground Loops. If it makes noise, it's a Ground Loop. If it doesn't make noise, it's a Ground Plane.

CHAPTER 22
ARE TUBES LOUDER THAN TRANSISTORS?

Does a tube amp sound louder than a transistor amp? To answer this question we need to study harmonic distortion, and the psychoacoutic effect.

DISTORTION

Any amplifier operated in its undistorted region will sound good. However, tubes and transistors have very different harmonic characteristics when driven into overload. These harmonics can be divided into three groups:

1. The second, fourth, and sixth (low order, even) harmonics produce a "choral," or "singing" quality.

2. The third, and fifth (low order, odd) harmonics produce a "covered," or "muffled" sound.

3. The seventh, ninth, and eleventh (high order, odd) harmonics add

"edge" giving the sound a sharp attack quality, and the illusion of "loudness."

The second and third harmonics are the most important. Musically, the second is an octave above the fundamental, and is almost inaudible, but has a doubling effect, adding body and fullness to the sound. The third harmonic will blanket the fundamental, and instead of adding fullness, actually makes the tone appear softer. A strong second can overcome the third, reopening the sound.

The "edge" harmonics (seventh, ninth, and eleventh) are musically unrelated, but if in balance can reinforce the fundamental, giving a sharp attack quality. Too much edge produces a raspy, dissonant quality. The ear is very sensitive to edge harmonics, and uses them to determine "loudness."

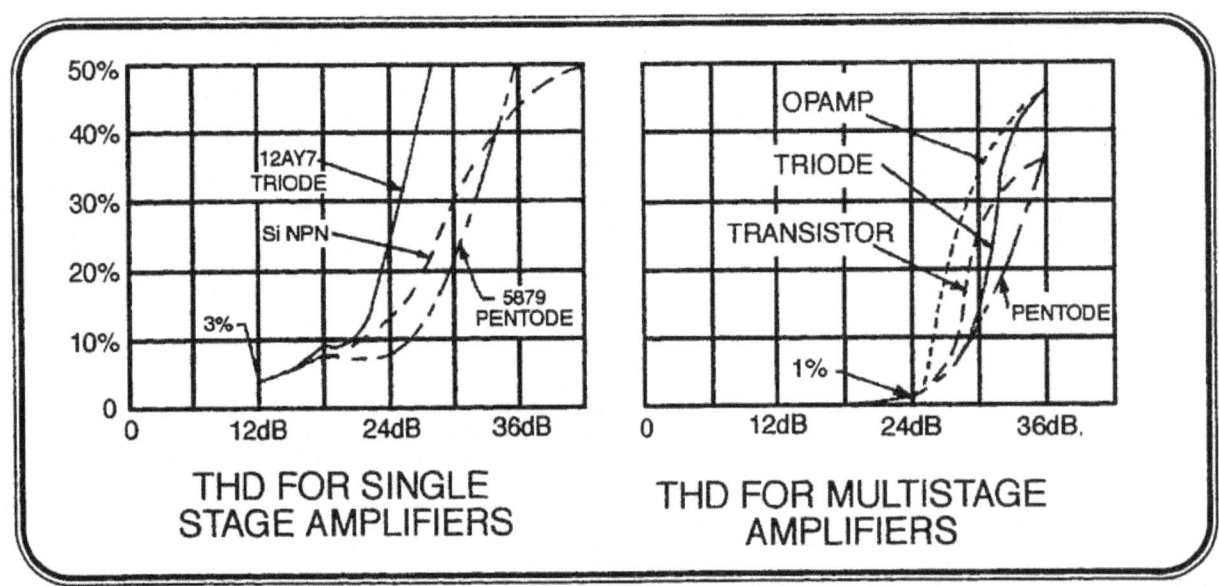

THD FOR SINGLE STAGE AMPLIFIERS

THD FOR MULTISTAGE AMPLIFIERS

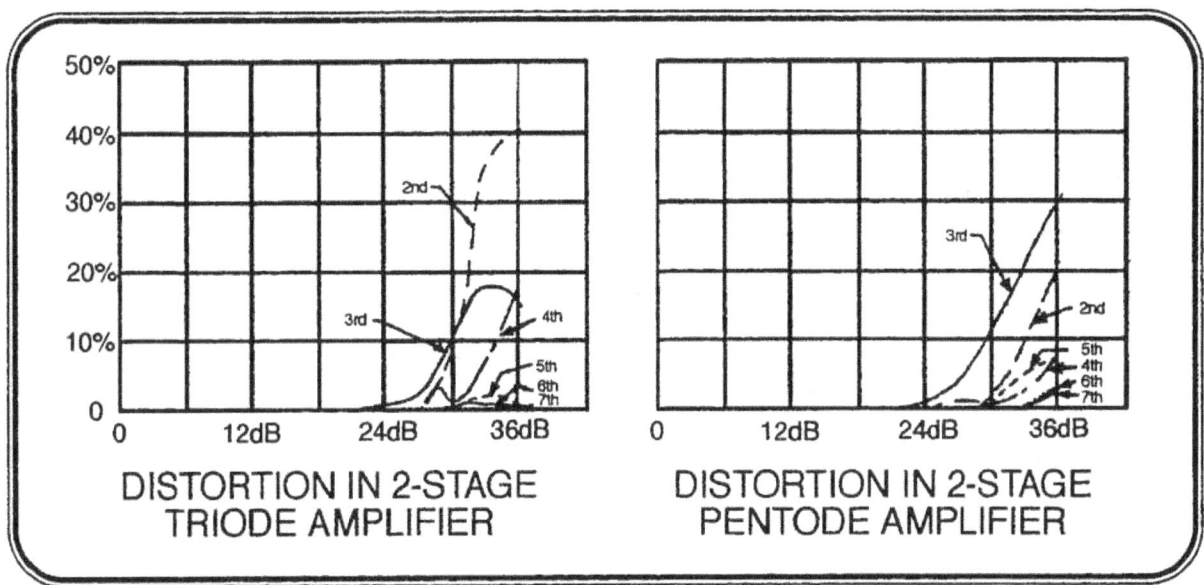

DISTORTION IN 2-STAGE TRIODE AMPLIFIER

DISTORTION IN 2-STAGE PENTODE AMPLIFIER

SOLID STATE AMPS

When solid state amps (including MOSFETs) overload, they contain a strong third harmonic, with a very weak second harmonic. The third harmonic produces a "covered" sound, giving the amp a restricted quality, and the illusion of a lack of power and punch, that the second harmonic is not strong enough to overcome. There is also a lack of "edge," or "loudness" harmonics at mild overload, again adding to the perception of a lack of power. Transistors only have about 10dB of distortion headroom.

VACUUM TUBE AMPS

An overloaded vacuum tube creates a whole spectrum of harmonics that are in balance with each other. The strongest are the second, then the third, fourth, and fifth harmonics, giving tubes a full-bodied, "brassy" quality. This gives tubes their "punch." Further into overdrive the amp produces a slow rising group of "edge" harmonics, which the ear translates as "loudness."

The combination of slow rising edge harmonics, with a predominantly "open" second harmonic structure forms an almost perfect compressor. In the top 15-20 dB of overload, the electrical output increases by only 2-4 dB, acting as a limiter. Since the "loudness" edge harmonics are increasing at the same time, the subjective loudness remains uncompressed to the ear, allowing tube amps to have a higher apparent level. Tubes sound louder, and have a greater signal to noise ratio, because of this subjective headroom effect. Tubes also reinforce an amp's bass response with their strong second and third harmonic content, reinforcing the "natural" bass with a kind of "synthetic" bass.

POWER MEASUREMENTS

Power is specified for an amount of distortion. I would not be surprised if guitar amps are measured at 5% distortion, or more. Of course, the power is higher at higher distortion levels, because you're pushing it further into clipping. Also, a 40-Watt amp can put out 80-Watts fully blown

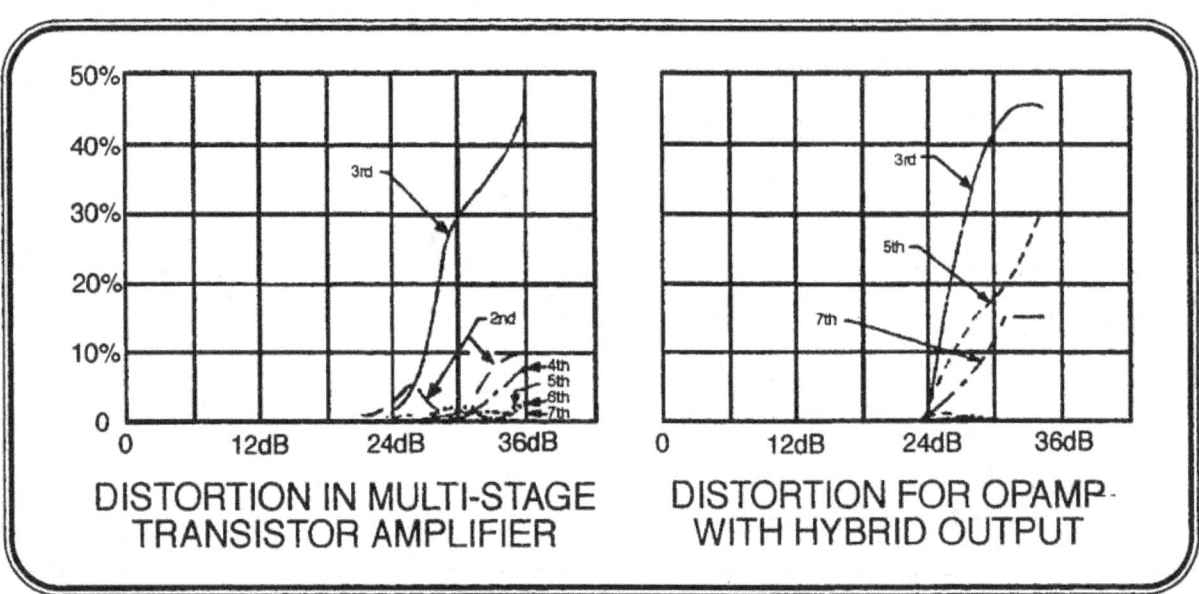

DISTORTION IN MULTI-STAGE TRANSISTOR AMPLIFIER

DISTORTION FOR OPAMP WITH HYBRID OUTPUT

(not clean). A Blackface or Silverface amp with two 6L6's puts out 40 Watts. A Twin or Showman puts out 81 Watts. In comparison, many "50-Watt" Marshalls put out 34 Watts, even though I hear all the time that 50-Watt Marshalls are LOUD. This perception of volume is based on a distorted output signal. The maximum "clean" power I've ever seen out of a set of four tubes was 112 Watts for MusicMan, Orange, Hiwatt, or Ampeg amps.

Transistor amps rarely have output transformers, or impedance selector switches. Transistor amps put out more power as the impedance of the attached speaker goes down because the lower speaker impedances more closely match the transistor's low internal impedance, resulting in a better transfer of power, and higher Wattages. For the maximum power out of a tube amp, you need to attach the right speaker load. As the speaker impedance goes down, you do not automatically get more power out of a TUBE amp.

Over the last few years, I have been telling people to think of a tube amp as a sound processor. Using one is like adding a phase shifter, flanger, or chorus to your signal. They can even make your singing sound better. Transistor amps just don't have these abilities, and is another reason that tube amps retain their immense popularity.

References: Tubes versus Transistors - Is There an Audible Difference, by Russell O. Hamm. JAES, May 1973, Volume 21, Number 4, pp. 267-273. Sear Sound Studios, New York, NY. Presented September 14, 1972, at the 43rd Convention of the Audio Engineering Society, New York.

NOTES